AF263025

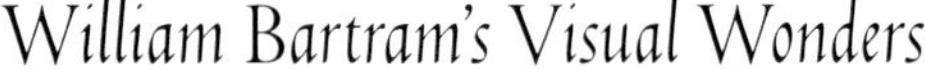

William Bartram's Visual Wonders

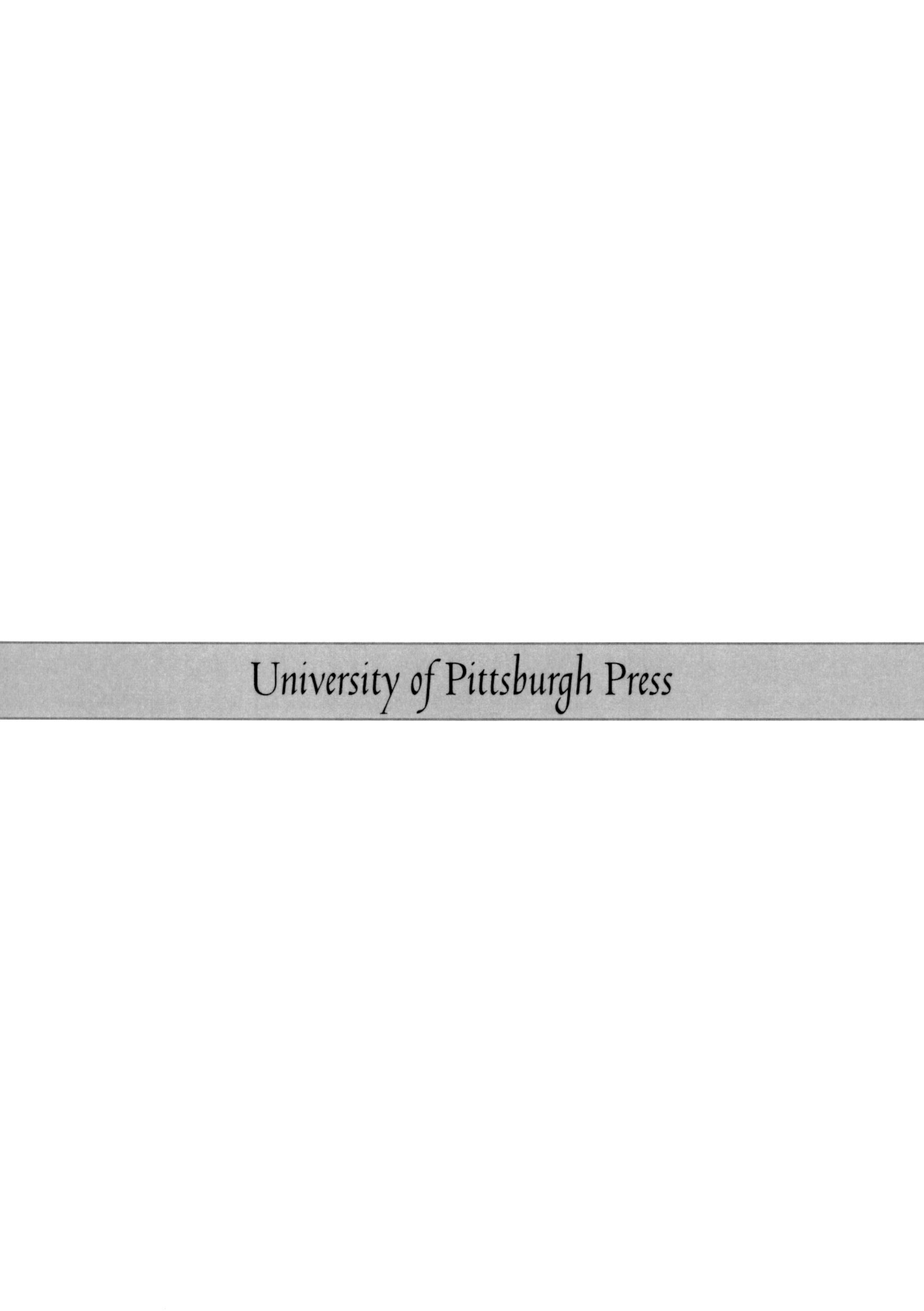
University of Pittsburgh Press

ELIZABETH A. ATHENS

WILLIAM BARTRAM'S
Visual Wonders

The Drawings of an American Naturalist

Published by the University of Pittsburgh Press, Pittsburgh, Pa., 15260
This paperback edition, Copyright © 2025, University of Pittsburgh Press
Copyright © 2024, University of Pittsburgh Press

Manufactured in the United States of America
Printed on acid-free paper
10 9 8 7 6 5 4 3 2 1

Cataloging-in-Publication data is available from the Library of Congress

ISBN 13: 978-0-8229-6761-3
ISBN 10: 0-8229-6761-8

COVER ART: William Bartram, *The Marsh Hawk from N:th America* [*Circus cyaneus* or northern harrier] *and the Reed Bird* [*Dolichonyx oryzivorus* or bobolink], *the Same with the Rice Bird of Catesby,* 1755–1756. Courtesy of Arader Galleries, New York; William Bartram, *Coach Whip Snake* [*Masticophis flagellum*] *from Et Florida,* 1774. © Natural History Museum, London.

COVER DESIGN: Alex Wolfe

Publisher: University of Pittsburgh Press, 7500 Thomas Blvd., 4th floor, Pittsburgh, PA 15260, United States, www.upittpress.org
EU Authorized Representative: Easy Access System Europe, Mustamäe tee 50, 10621 Tallinn, Estonia, gpsr.requests@easproject.com

For my parents

Contents

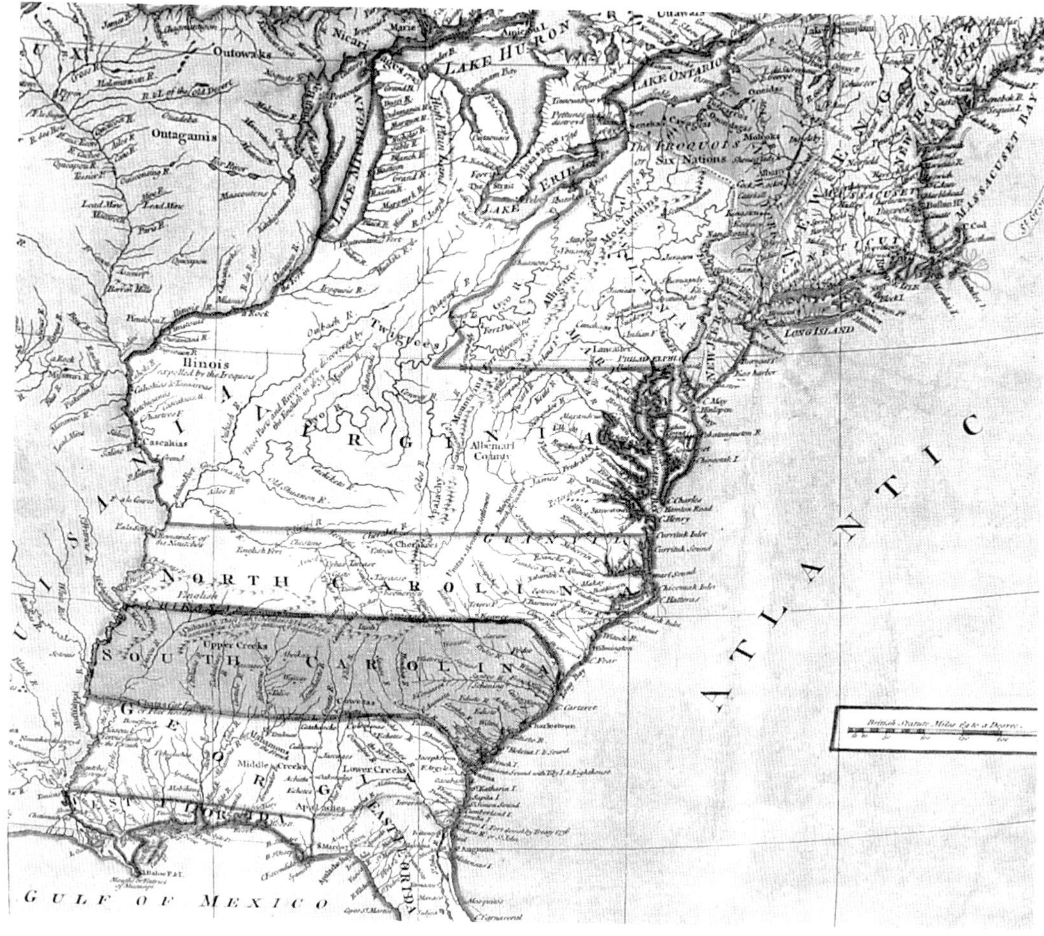

FRONTISPIECE. Bell's 1771 map identifies key sites in William Bartram's life, which include the city of Philadelphia; the Cape Fear River in North Carolina, where Bartram lived with his uncle and ran a shop between 1761 and 1765; Charleston, South Carolina, a starting point for his botanizing expeditions in 1765 and 1773; and the St. Johns River in East Florida, where he attempted to establish a plantation in 1766. Peter Bell after Jean Baptiste Bourguignon d'Anville (Richard Seale, engraver), *A New and Accurate Map of North America* (London, 1771), detail. Geography and Map Division, Library of Congress.

Acknowledgments

In a book dedicated to the construction of knowledge, it is only right to begin with thanks to those who made this project possible. Deserving of my profound gratitude is Amy Meyers, whose scholarship has provided the very foundation for the critical study of William Bartram's drawings. For much of her career she has served as the director of the Yale Center for British Art, and her professional life has been devoted to ensuring that others have the resources needed to develop their scholarship. Even while aiding an extensive transatlantic network of curators and researchers, she has produced a stream of essays and lectures offering new insights into the art and working relationships of the Bartrams and their contemporaries. I was fortunate to attend two gatherings she co-organized with Therese O'Malley in 2018—at the Oak Spring Garden Foundation in Virginia and at the Paul Mellon Centre in London—during which I had the opportunity to speak with colleagues whose work on botanical exchange, the history of gardens, and the visual culture of natural history has deeply influenced my own. Amy's great kindness and scholarly generosity are matched only by her encyclopedic knowledge, and it is my tremendous privilege to have her as a mentor and friend.

I would also like to pay tribute to Joel Fry, the longtime curator of Bartram's Garden. His knowledge of the Bartrams and the world of early American botany ran deep, as did his willingness to discuss these topics with amateur enthusiasts and devoted scholars alike. Most valuable was Joel's ability to research John and William Bartram with single-minded intensity, while also recognizing the gap between their day and ours. William especially is a figure with whom contemporary scholars identify, making it tempting to replace his views with modern-day perspectives, but Joel conscientiously guarded against this. I was fortunate to have him read a draft manuscript of this book before his death in the spring of 2023, and his comments and corrections were invaluable. Joel's absence leaves a terrible void in Bartram studies.

Institutional support, too, was critical to the book's development. I thank the UConn Humanities Institute (UCHI) for a 2020–2021 fellowship that allowed me time to work on the manuscript. Alexis Boylan, Yohei Igarashi, and Michael Lynch—having to adapt UCHI's fellowship program to accommodate the coronavirus pandemic—crafted a brilliant community of writers and researchers. My colleague Helen Rozwadowski served as an important interlocutor that year. Beyond UCHI, my work has also been supported by the American Philosophical Society, the American Council of Learned Societies/Henry Luce Foundation, the American Society for Eighteenth-Century Studies, Arader Galleries, the Library Company of Philadelphia, the Paul Mellon Centre, and the Yale Center for British Art.

Additional support came from a residency at the Clark Art Institute in Williamstown, Massachusetts, where I found a summer idyll in the Berkshires and the opportunity to discuss my research with senior scholars, including Christopher Riopelle of the National Gallery in London, Andrés Úbeda de los Cobos of the Museo Nacional del Prado, and Peter Samis of the San Francisco Museum of Modern Art. My gratitude goes to them and to the museum's research and academic programs staff. I also had the good fortune to work as a postdoctoral research associate for Therese O'Malley at the Center for the Advanced Study of the Visual Arts at the National Gallery of Art in Washington, DC, which allowed me to benefit from her extensive knowledge of historical American landscapes. I thank Tim Barringer, Lisa Ford, Michael Gaudio, Jim Green, Andrea Hart, Judith Magee, Alex Nemerov, Alison Petretti, Jenny Raab, and Ann Upton for their guidance and assistance. Abby Collier of the University of Pittsburgh Press has been entirely gracious during the process of developing this book, and I thank her, her team, and the manuscript's reviewers for their thoughtful suggestions.

The materials in chapters 2 and 4 were first presented in various guises at the University of Leeds and at the McNeil Center for Early American Studies in Philadelphia. My thanks go to Lara Eggleton and Richard Checketts, who together developed a fascinating conference on ornament at Leeds, and to Mairin Odle, who coordinated my panel at the McNeil Center. Discussions with fellow scholars at both venues helped shape my approach to ornament, natural history, and the theorization of drawing during the eighteenth century. I was able to publish short essays based on these presentations in the *Journal of Florida Studies* and in the edited volume *The Attention of a Traveller: Essays on*

William Bartram's Travels *and Legacy*, put out by University of Alabama Press. I am grateful to both Thomas Hallock and Kathryn E. Braund, who invited me to contribute to these publications. My thoughts on the overlap between William Hogarth's *The Analysis of Beauty* and the practice of natural history and natural philosophy—so crucial to my understanding of Bartram's drawings—were first worked out in special issues of *J18: A Journal of Eighteenth-Century Art and Culture* and *Oxford Art Journal*; I thank Meredith Martin, Noémie Etienne, and Katie Scott for those opportunities.

Friends, colleagues, and family merit special notice. Xiao Situ and Eloise Patterson were invaluable supports during our PhD program and after; I have benefited enormously from their kindness and insight. I am also grateful to my unofficial advisor and official friend Marc Simpson, who has always been ready with sound advice. Equally deserving of mention are Christine Paglia-Baker and Nick Baker, Kelly Dennis, Robin Greeley and Michael Orwicz, Yoko Hara and Junya Mikami, Chloe Kiritz and Matt Melonio, Kathryn Moore, and Macushla Robinson. To my family—Bill Sr., Bill Jr., Brenda, Jessica, Jules, and Owen—I send much love. As always, my deep gratitude, admiration, and love go to Sam.

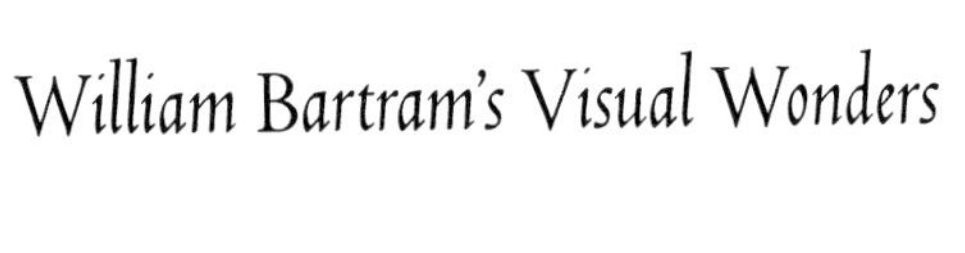

William Bartram's Visual Wonders

Introduction

Science's Brightest Ornament

Died, suddenly, at Philadelphia, on July 22d, by a rupture of a blood vessel, Mr. WILLIAM BARTRAM, senior, in the 85th year of his age. By his death, science has lost one of her brightest ornaments.

—*Washington Quarterly Magazine of Arts,*
Science and Literature (Apr. 1, 1824)

This succinct summary of William Bartram, printed nearly nine months after his death, points to his crucial role in what we now understand as modern science. Born in rural Philadelphia County in 1739, he contributed to many fields of study over the course of his long life. He collected plant and animal specimens as far north as New York and as far south as East and West Florida, colonies that encompassed parts of modern-day Florida, Alabama, Mississippi, and Louisiana. He provided some of the first published accounts of Native communities in the Southeast, wrote on the migratory patterns of swallows and other birds of passage, described the hybridization of grape vines in North America, and was an important interlocutor for many of the most noted naturalists of his day. His interests and abilities were manifold and his network of colleagues vast.

Today he is perhaps best known as the author of a travelogue describing a botanizing journey he made through the American South between March 1773 and January 1777. Published in 1791, his *Travels Through North & South Carolina, Georgia, East & West Florida, the Cherokee Country, the Extensive Territories of the Muscogulges, or Creek Confederacy, and the Country of the Chactaws;*

Containing An Account of the Soil and Natural Productions of Those Regions, Together with Observations on the Manners of the Indians has almost come to stand in for Bartram himself. The book—part natural history, part religious allegory—has been the primary lens through which we now view Bartram's contributions to science and is the starting point for biographical studies of the naturalist. In some ways, this emphasis is not surprising: soon after the book's publication in Philadelphia, it was reprinted in London and Dublin and was translated and published in Amsterdam, Berlin, and Paris. Bartram's *Travels* became an especially influential text to the Romantic poets, who were drawn to the plash and flow of his language.[1]

Writing was not, however, Bartram's only or even preferred method of recording the world around him. Drawing was by all accounts his "darling delight," even though most of his illustrations remained unpublished.[2] Because they were circulated as bespoke, semi-private renderings among a coterie of fellow naturalists, his drawings have attracted less scholarly attention than his written work. And yet, drawing was central for Bartram: it was at once a means of discovery, a form of cognition, and an act of imaginative collaboration and memory. He viewed drawing as a series of reciprocal interactions and interdependencies among the natural world, the artist naturalist, and the observer, and this understanding deeply influenced his view of a dynamic and mutable cosmos. With this book, I trace the roots of Bartram's natural history to his drawing practice.

William was raised from a young age to attend to the natural world. He was the fourth son of John Bartram, a Pennsylvania farmer, self-educated naturalist, and the most important purveyor of North American botanicals to Britain in the eighteenth century. John made numerous botanizing trips through the countryside near his home located in Kingsessing, just west of Philadelphia, and traveled throughout Britain's American colonies, collecting plants and seeds. These he sent to colleagues abroad, who distributed them among British nurseries and gardens. By his early teens William had become his father's preferred companion on such collecting expeditions, and he began making natural history drawings to accompany these shipments.[3] An early example of his work is a plan of the Bartram family's botanic garden, which identifies John's study, the garden's various plots and walkways, as well as its central pond.[4] The drawing, informative as it is, also incorporates elements of William's

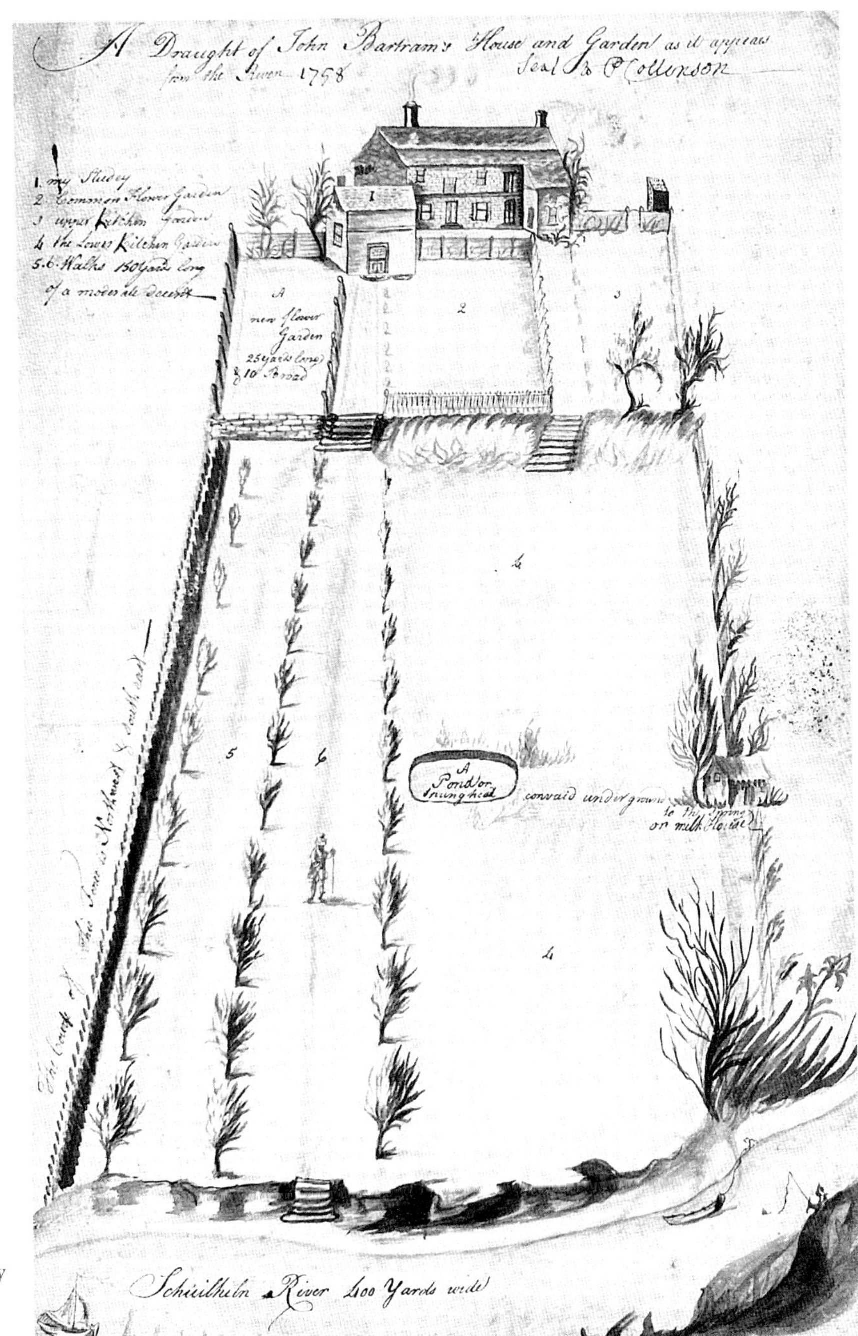

Fig. I.1. William Bartram, *A Draught of John Bartram's House and Garden as It Appears from the River*, 1758. Courtesy of Arader Galleries, New York.

telltale whimsy. The Schuylkill River, which borders the eastern edge of the garden, shown at the bottom of the drawing, is dotted with two small vessels and a fisherman. A man with a walking stick—possibly John—looks out at the riverway (fig. I.1).

John Bartram recognized his son's aptitude in botany and drawing, but he saw it as a pastime, not a profession. In 1755 he wrote of his concerns to his most important patron, the London mercer Peter Collinson, who served as a hub for the vast international exchange of natural history materials in the eighteenth century. "My son William is just turned of sixteen, [and] . . . it is now time to propose some way for him to get his liveing by," he observed. Although the younger Bartram had already been admitted to the Academy of Philadelphia, John was not keen on him becoming a "gentleman." Rather, he explained to Collinson, "I want to put him to some business by which he may with care & industry get A temperate resonable liveing I am afraid Botany & drawing will not afford him one."[5] After considering a number of steady and respectable vocations for his son, including printer and engraver, John apprenticed William to the Philadelphia merchant Captain James Child in January 1756.[6]

By all accounts, William was industrious in his apprenticeship with Captain Child, but he struggled in his subsequent attempt to establish himself as a merchant at Ashwood, North Carolina, where he lived with his uncle Colonel William Bartram and his family. Letters from that time speak to the difficulty he experienced. "I am unfortunate in ariving to a bad Markett, a wrong Season of the year, and the excessive rains has almost destroyed the Country," he informed his father in May 1761.[7] Although John often inquired after William's "success in merchandise," he also wrote with many requests for seeds, plants, and other observations of the natural world.[8] From the smattering of letters that remain it seems William devoted as much of his time to botanizing as he did to his business endeavors. He struggled to turn a profit and was plagued with debts. In 1764 he explained to his older half brother Isaac, "I would write to Father but I am afraid my Letters gives him Uneasiness." He was clearly more comfortable reflecting instead on the natural world. "When I view the eternal necessary actions and movements of the Viseble Universe," he wrote in the same letter, and "the amaising exactness and truth in the effects of every cause in the progress of Nature, I see an ab[s]alute necessity for the constant activity in every Object & part of the creation."[9] In his mature graphic work William seemed intent on forging a connection between the operations of his pencil and the operations he observed in nature.

In 1765 John Bartram was recognized with a royal pension to explore East Florida, which had recently become part of the British imperium. Styling himself the "King's Botanist," he invited William to join him on an expedition along the St. Johns River.[10] Likely relieved to put aside his merchant business, William asked his uncle to settle his outstanding debts, and he accompanied his father in search of flora and fauna. During this trip he became so taken with the region's dramatic landscape that, much to his father's chagrin, he decided to establish an indigo and rice plantation near St. Augustine. Despite misgivings, John purchased tools, seed, and six slaves for William's plantation, which met an even more ignoble end than his merchant business. The acres William acquired were swampy and stagnant, the weather was unbearably hot, and he was plagued by illness, despair, and most likely, profound guilt for a speculation premised on slave labor. Henry Laurens, a South Carolina plantation owner and slave trader who had advised the Bartrams, visited William in the summer of 1766 and wrote to John of William's "forlorn state." Laurens suggested John send additional provisions, but John demurred, instead encouraging William to return home.[11]

Bartram's years as a merchant and planter interfered with his drawing, but by 1767 he was wholly recommitted to the practice, and he remained so until age and failing eyesight forbade it. He again made figures of biota for Collinson and also began sending drawings to the London physician John Fothergill, who agreed to underwrite a nearly four-year botanizing journey through the Carolinas, Georgia, and East and West Florida. During this 1773–1777 excursion, Bartram sent Fothergill seeds and specimens of subtropical plants, along with numerous drawings and field notes, which would later form the basis of *Travels*.[12] Yet even as Bartram dedicated himself to this writing project, one that took nearly fifteen years to complete, he continued to make visual records of the natural world for himself and for others. In 1788, for instance, he prepared a sheaf of watercolors for the London-based brewer and amateur botanist Robert Barclay, and in the 1790s he designed the illustrations for Benjamin Smith Barton's 1803 *Elements of Botany*, the first American botany textbook. From his early teenage years until deep into his sixties, Bartram turned to drawing first and foremost to understand and interpret the bewildering beauty of the organic world. The very act of drawing, of figuring a surface, has always lain at the heart of William Bartram's understanding of nature.

To Fi'gure. v.a. [*figuro*, Latin.]

But what does it mean—or what did it mean—for William Bartram to figure the natural world? What were the historical valences of the verb *to figure*, and how might they have informed his practice? To gain a sense of how it was conceived and understood, Samuel Johnson's 1755 dictionary offers some guidance. *To figure* could mean to form something into a definitive shape, to approximate a visual or physical resemblance, to ornament, to diversify, to imagine, or to represent by metaphor or visual symbol. This semantic range seems so bewilderingly broad as to encompass opposing meanings. In its most basic, primary sense, *to figure* could mean simply to manipulate matter, to engage in some sort of direct intervention in the physical world. In a secondary, representational sense, it followed two diverging trajectories: on the one hand, it could mean the creation of an uncomplicated correspondence between a thing and its textual or visual representation, the forging of a likeness, an immediate and obvious concordance. On the other hand, it could also suggest a misalignment, a sideways step toward the object one intended to describe. Johnson cited John Locke's "On the Conduct of the Understanding" to explain this type of representation: "Figured and metaphorical expressions do well to illustrate more abstruse and unfamiliar ideas, which the mind is not yet thoroughly accustomed to."[13] In this sense, figuring is allusive or even ornamental; a figure gestures toward an object or an idea, while at the same time underscoring the fact that it remains separate and distinct from that idea, a type of representation that calls attention to its own mediation. The various definitions of figuring move by degrees away from matter toward thought and imagination.

The semantic range of *to figure* makes it an inherently unstable term. In his graphic work Bartram seemed to embrace its multiple meanings, drawing together its material, literal, and figurative connotations. He did not pit direct empirical observation against fancy, ornament, or allusion but, instead, knit them closely together. His visual representations of the natural world function as a graphic intermediary between material manipulation, sensory experience, and mental construct, delineating nature as he experienced and came to understand it.[14]

Bartram's collapse of the various meanings of *to figure* may be unusual, but it was not without precedent in the eighteenth century. The most likely influence on his conception of figuring comes from the artist and engraver William Hogarth, then one of the most celebrated artists in the English-speaking world. In his well-known treatise, *The Analysis of Beauty* (1753), Hogarth worked to dissolve the inherent opposition between the literal and the allusive by joining them together in his Lines of Beauty and Grace, which he framed as the very basis of his theory of representation. Hogarth found these serpentine and coiling lines best suited to representing the living world because of their fluidity and flexibility; such qualities, he believed, allowed them to express the greatest diversity of forms. At the same time, Hogarth also considered them to be the most beautiful and aesthetically pleasing marks because of their ornamental qualities. The way they spool across the surface of the paper leads the eye in pleasurable pursuit and suggests ongoing, unfolding transformation over time. The Lines of Beauty and Grace served in Hogarth's representational model as the bass figures for both truth and beauty, essence and ornament. Hogarth's *Analysis* was circulating in Philadelphia by the 1760s, and it seems from Bartram's compositions that he came to share the older artist's view of figuring as a process that was at once manual and intellectual, literal and allusive.

Naturalists who have engaged with Bartram's natural history, both in his day and ours, have often attempted to tease apart the literal from the metaphorical, hoping to pierce through the veil of poesis to access the truth of his observations. The title of Berta Grattan Lee's 1972 essay, "William Bartram: Naturalist or 'Poet?,'" highlights how profoundly this question has animated the study of Bartram and his work. As early as 1767, Bartram's future patron John Fothergill noted his skill in drawing but lamented his tendency to include "Imaginary Plants" in his compositions.[15] Notably, the biota Fothergill was referring to was not, in fact, imaginary; he took issue with Bartram's portrayal of a Venus flytrap (*Dionaea muscipula*), a species then little known in Europe. Although Fothergill was one of the first to question the veracity of Bartram's observations, it was not the last. Naturalists and nurserymen followed his footsteps through the American South, endeavoring to ascertain what was fact and what was fancy in Bartram's work. The urgency to parse the accurate from the imaginative is pervasive throughout the Bartram scholarship. Francis Harper—the enterprising scholar who laid the foundation for Bartram research in the twentieth and twenty-first centuries—was largely motivated by this desire. Trained as a biologist,

Harper devoted much of his career to retracing Bartram's steps and determining the chronology of his travels. His copiously annotated naturalist's edition of Bartram's *Travels*, published in 1958, represents the summa of these efforts.[16]

Yet for Bartram the literal and metaphorical, ground and figure, were always inseparable, interpenetrating fully in his work. The way he negotiated and ultimately collapsed notions of figuring reinforces the importance of drawing to his understanding of nature. As the creation of marks on a surface, drawing provided a direct manipulation of the world (figuring in its primary sense) as well as a literal and metaphorical representation of it (figuring in its secondary senses). Even so, the majority of Bartram scholars in the humanities have focused on his writings, rather than his graphic enterprise. Literary historians were the first to engage with Bartram's work, as we see in John Livingston Lowes's 1927 *The Road to Xanadu: A Study in the Ways of the Imagination*. In his analysis of Samuel Coleridge's poetry, Lowes argues that Coleridge pulled from Bartram's descriptions of the Florida landscape to create the heady, disorienting atmosphere in *Kubla Khan*. N. Bryllion Fagin, turning from *Travels* as a source of literary inspiration to a piece of literature in its own right, published *William Bartram: Interpreter of the American Landscape* in 1933. He argues that Bartram entwined accurate observation with subjective interpretation to offer a rich, multisensorial experience of nature.

Many more in-depth examinations of Bartram's writings were to follow: Thomas Hallock, Nancy Hoffmann, Christopher Looby, Pamela Regis, and others have all ably explored the literary origins and influences, style, and expressive affect of Bartram's *Travels* as a text. Scholars in other fields have used Bartram's *Travels*, correspondence, and commonplace books to examine his engagement with different social, political, and cultural questions, including the American Revolution, Native American policy in the colonial period and the early Republic, and the practice of slavery. These projects reveal the critical role he plays in different disciplines, yet they derive almost exclusively from what Bartram wrote, not from what he drew. If, as John Bartram declared, drawing and botany were William's primary interest and pleasure, then it seems apt to turn our attention to his graphic work and to what it can teach us about his approach to and understanding of the natural world.

There exists a small but important body of scholarship dedicated to his drawings, developed by historians of science Joseph Ewan and Judith Magee, as well as art historians Amy Meyers and Michael Gaudio.[17] In particular, Meyers's

1985 doctoral dissertation, "Sketches from the Wilderness: Changing Conceptions of Nature in American Natural History Illustration, 1680–1880," established the terms by which Bartram's drawings are now considered. She reads his work as part of a centuries-long shift in which nature is increasingly viewed as mutable and interdependent as opposed to hierarchically ordered. This shift, she argues, is more clearly articulated in Bartram's graphic work than in his *Travels*; through formal repetition and the use of an animated, agitated line, his images reinforce the interdependencies and dynamic relationships among organisms, even when his text adheres to a more categorical view of the cosmos.[18]

Building on Meyers's scholarship, Gaudio examines how a mutable natural world forces organisms into a continual "struggle for self-evidence." Bartram understood the new republic to work by principles similar to those in nature, Gaudio observes, and Bartram's drawings are as much about the social order of the new nation as they are about the structure of the natural world.[19] Like Gaudio, I build on Meyers's analyses, but rather than expand her discussion of Bartram's view of the cosmos into the social realm, I instead trace it back to its origins in his mental and manual acts of figuring.

This emphasis on drawing as a means of coming to know and understand nature is, I argue, what distinguishes Bartram from his colleagues. Even though other naturalists at the end of the eighteenth century had also begun to see the natural world as mutable and interdependent, this conception of nature did not necessarily align with their graphic practices. Both Erasmus Darwin and Jean-Baptiste Lamarck developed early theories of evolution, proposing the idea that all organisms were not wholly separate and distinct but, rather, emerged from a single common ancestor. Lamarck in particular identified this evolution as a response to environmental conditions, though unlike Charles Darwin (the grandson of Erasmus), he thought evolution happened within an organism's lifetime rather than over generations through natural selection. Lamarck's views regarding environmental influence were taken up in part by the enterprising explorer and naturalist Alexander von Humboldt, who introduced the concept of plant geography; that is, the natural distribution of plants according to latitude and altitude. Humboldt did make sketches during his botanizing expeditions and encouraged other artists to delineate the complexity of the material universe, yet much of his graphic work lacks the dynamic, vivifying quality of Bartram's most curious compositions.

Although my approach to Bartram's graphic practice underscores the absolute centrality of drawing to his natural history, this is not to say that *Travels* goes unaddressed in this study. The drawings serve as the main source of evidence and point of departure, but I occasionally pivot to his text to corroborate interpretations. It is essential to remain cognizant that *Travels* is a work of revision and rewriting and that Bartram's voice is at least partly shaped by editorial intervention, as Nancy Hoffmann has shown.[20] Still, it is an important component of Bartram's natural history and, more specifically, his legacy as an artist naturalist. The book has been important in shaping ideas about Bartram and his work since its publication, when it was quickly put to use as a field guide for botanizing expeditions in the American South. *Travels* ultimately aided in the process of remembering and "re-figuring" Bartram's natural history, as later generations of naturalists and nurserymen retraced his mazy path. But far before this text—before its drafting, its revisions, publication, translations, and new editions—there were Bartram's drawings.

Chapter Overview

I begin this study with an examination of how John Bartram, a farmer with no formal training, became a key figure in the transatlantic trade in plants, introducing his son to the study of nature and to the ocean-spanning coterie of virtuosi who would underwrite that study. More specifically, in chapter 1 I consider the asymmetry of access between Europeans and Americans (access to European scholarly infrastructure on one hand and the American wilderness on the other) and the role it played in the development of William's natural history. As Susan Scott Parrish points out, the relationship he and his father enjoyed with their patrons abroad was never fully equal. The Bartrams provided keenly sought specimens of an unfamiliar and otherwise inaccessible natural world, and their patrons offered access to resources and scientific communities absent in the colonies. Yet European naturalists often viewed their American correspondents as unsophisticated countryfolk, incapable of making sense of the wilderness they inhabited. Only in Europe could the natural fragments provided by colonials—roots packed in muck and moss, dried fish, bird skins preserved in tobacco or spirits, fragmented descriptions, and field sketches—be synthesized and rendered intelligible.

Both father and son bristled at the mantle of preeminence their European colleagues assumed, and they worked to establish themselves as intellectual equals. John's most concerted effort in shifting this imbalance was to craft his own taxonomy of North American trees between 1753 and 1756, which he supplemented with drawings by William. Both his system and William's illustrations were made, he claimed, "not according to science or art, but nature," offering the truest and most accurate representation of the natural world they inhabited.[21] He sent these materials to Peter Collinson in London but became defensive when Collinson suggested that John failed to "distinguish things aright."[22] Unmoved, John asked Collinson to submit to him any errors for correction and pressed him to find an engraver and printer for his classification system.[23] Collinson pursued various avenues, but by 1756 the project had stalled. Collinson was comfortable with the Bartrams' collection and preparation of American flora and fauna but clearly balked at their attempts to articulate a system for understanding and interpreting them.

Although the whereabouts of much of John's arboreal taxonomy and his son's associated drawings are today unknown, a 1767 drawing of an American lotus (*Nelumbo lutea*) by William Bartram (the subject of chapter 2) offers a clear sense of his mature notion of figuring. This was the drawing that Collinson used to convince Fothergill to support William's botanizing, and the one that elicited Fothergill's perplexity as well as his admiration. Not only did Bartram feature a Venus flytrap, which Fothergill assumed to be invention, but he failed to hew closely to period standards of natural history representation. Although Bartram delineated the lotus's leaf, flower, and seed as was standard for the time, he also depicted the blossoms towering overhead, dwarfing the great blue heron in the foreground, and he flattened one of the lotus leaves, as though it were a pressed specimen. The inconsistent scale and multiple perspectives, along with the drawing's inclusion of a nondescript species, are strange and unruly, suggesting the idiosyncratic work of an untrained novice. In this chapter I argue that, far from amateurish, Bartram's graphic output pointedly responds to the power dynamic in eighteenth-century transatlantic natural knowledge in an incredibly sophisticated way. In providing his disruptive renderings to European virtuosi, Bartram called attention to his own processes of exploration, discovery, and knowledge formation.

In this work, Bartram visually quoted the illustrations of the noted ornithologist George Edwards, pulled from recent publications on plant systematics, and structured the drawing in accordance with William Hogarth's edicts on proper composition. The citations reference epistolary exchanges and published textual and visual sources that would have been known to the drawing's viewers, conveying to them Bartram's understanding of empirical science, his study of natural history, and his mastery of descriptive botany. These and other references are connected through a series of formal rhymes that lead viewers through the drawing's different stations, allowing them to reexperience Bartram's own educational process. Bartram's drawing is a visual manifesto that articulates his notion of figuring. The drawing does not fail to accord with established conventions so much as it registers the failure of the conventions themselves.

Not surprisingly, Bartram's drawings often resist not only the representational conventions of eighteenth-century natural history but also its taxonomic imperatives. By the 1730s Carl Linnaeus, the most celebrated systematist of the period, had established a rigid organizational scheme that arranged nature into kingdoms, classes, orders, genera, and species according to their ostensibly immutable essences. Such a system, however, offered little space for ambiguity and no sense of the natural world's lived relationships. In chapter 3, I consider how Bartram—influenced by his empirical observations as well as by his engagement with microscopy and his appreciation of the emerging theory of vital materialism—subtly challenged Linnaeus's taxonomic model.[24] The shifts in his graphic work between the massive and the minute along with his focus on hybrid forms draw on the defamiliarizing effects of the magnified view and the leveling perspective of vitalistic thought.

The practice of microscopy and the theory of vital materialism were closely intertwined. Microscopy dramatically unhinged both scale and perspective and subverted the sense that the natural world was visually self-evident. In one of the texts owned by the Bartram family, the author queried with awe, "Who, a thousand Years ago, would have imagined it possible, to distinguish Myriads of living Creatures in a single Drop of Water? . . . Or, that numberless Species of Creatures should be made visible, tho' so minute, that a Million of them are less than a Grain of Sand?"[25] Others claimed that such studies revealed how all organic matter, even when separated from the organism, possessed a vital, living force.[26] This view challenged the basis of Linnaeus's taxonomic project by suggesting a radical underlying sameness among living beings, a sameness that made the boundaries between species, genera, and even kingdoms porous. These ambiguities made many naturalists uneasy, but they seem to be precisely what drove Bartram's investigations. In this chapter I focus on drawings he made for Fothergill, in which he presented unexpected shifts in scale, as well as jittering, agitated forms and hybrid creatures. By conveying the instability and complexity of the natural world in drawings such as these, Bartram interrogated artificial categorization as the basis of natural knowledge and instead proposed a different kind of knowing, one that embraced the disruptiveness of living nature and its metamorphic processes.

In his drawings Bartram aimed to reveal his processes of exploration and discovery as well as the rich complexity of the living world, yet these aims presented a representational and ideological challenge. In a mutable natural world, there exists no essential appearance of a specimen, which means that representation is always in danger of denervating the portrayed organism in a manner that runs counter to lived experience. In chapter 4, I examine how Bartram offset the potential petrifying effects of representation through a variety of means, most markedly through his reliance on drawing. According to David Rosand, drawing possesses an especially "open" graphic structure in which figure and ground retain their autonomy, revealing the conditions and gestures of the composition's construction. The representational ambivalence of drawing—its ability to refer back to the artist and out toward the world—allows for a more dynamic image. Such instability in meaning demands the viewers' participation and imaginative engagement, and it is through this collaboration—in effect, the artist's and the viewers' joint re-creation of the world—that Bartram's graphic work gathers a sense of liveliness and vitality.

Caroline van Eck calls this effect the "living presence response," which she examines in the context of eighteenth-century sculpture. Although her research is centered on traditional "high" art, her study of a form of representation so vivid that it appears either to possess agency or to transmute into the object it portrays applies equally to Bartram's drawings. Van Eck traces the origins of living presence response to classical rhetoric and its theorization of *figura*, which she understands as the modification of text or image from its standard representational use. *Figura*, she observes, "can refer to all expressions that have 'received a new aspect.'" This type of figuring prompts the viewers' experiential memory and affective response, enlivening the representation.[27] As a student of classics at the Academy of Philadelphia, Bartram would have been familiar with Quintillian, Cicero, and the other rhetoricians whom van Eck cites. Taking as a central focus a number of drawings Bartram produced for Fothergill during his expedition, in chapter 4 I examine how Bartram endeavored to create images of flora and fauna that were as lively and dynamic as nature itself.

The years between 1767 and 1777 marked the most original period of William Bartram's work as an artist naturalist, as he actively *figured out* his understanding of the natural world. In January 1777 after he returned to Pennsylvania from his southern travels, he actively set out to shore up his and his father's legacy. He reviewed travel journals, specimens, and drawings, recombining them in ways that he hoped might introduce his investigations and insights to a broader public. Despite his robust correspondence with naturalists at home and abroad, he had yet to be identified for any of his botanical discoveries. In chapter 5 I examine his efforts to establish his name in the growing field of international botany. I consider a suite of four watercolor drawings, purged of most of the idiosyncrasies typical to his oeuvre, that he made in 1788 for Robert Barclay. I also consider the importance of his role as mentor to a younger generation of naturalists, especially as his waning eyesight made drawing impracticable.

In 1788 Barclay was a member of the newly founded Linnean Society, which, along with the Royal Society, had taken on the responsibility of identifying and naming new plants following the death of Carl Linnaeus. He was also much involved in the development of the *Botanical Magazine*, a popular illustrated periodical featuring exotic ornamentals, and in the compiling of the *Hortus Kewensis* (1789), a catalogue of the flora grown at the Royal Botanic Gardens, Kew. Bartram likely saw Barclay as an advocate for his family, ensuring that their service to botany received institutional acknowledgment. In their correspondence, Bartram asked only that he and his father be recognized as the discoverers of these American plants featured in the watercolors. In particular, he desired that a "very elegant flowering Tree" he found in Georgia and East Florida be named *Bartramia* in honor of his father, John, whose work so richly contributed to botanical knowledge.[28] The plant's current scientific name, *Pinckneya bracteata* (Georgia fevertree), reveals that this wish would not be realized. John Bartram had only a handful of mentions in the first edition of the *Hortus Kewensis,* and William none at all.

Although the method of establishing his legacy through institutions was disappointingly unproductive in some respects, Bartram's acts of mentorship and the shared experience of place proved more effective. Even as he was failing to gain recognition in British circles for his botanical discoveries, he was also shepherding a new generation of American naturalists along in their studies. Benjamin Smith Barton, the professor of botany and materia medica at the University of Pennsylvania was among them. He used the Bartram family garden as a classroom and collaborated with William on the development of the first

American botany textbook. Other naturalists took to the field, retracing William's path through the American South and collecting specimens of plants he had previously discovered. I conclude this chapter with an examination of the botanizing travels undertaken by André Michaux, author of the *Flora boreali-americana* (1803); Moses Marshall, of the Marshallton botanic garden in Chester County, Pennsylvania; and John Lyon, plant collector for The Woodlands estate outside of Philadelphia, all of whom retraced in part Bartram's southern expedition, engaging in literal recollections of his journeys. These more personal and affective modes of legacy production helped in establishing a history of Bartram's botanical contributions that was as rich and dynamic as his own understanding of the natural world.

The compound nature of figuring, a process at once corporeal, intellectual, and imaginative, inherently yields both overlaps and fissures. We see this process play out in Bartram's oeuvre: some themes or subjects he revisited regularly in his drawings and writings, retracing and reworking them almost compulsively, while to others he barely gave passing notice. His work often balances uneasily between over-description and under-description, contrasting detailed and thickly rendered passages with ones that are sparse and only faintly sketched. In this book I follow a similar pattern, with areas of concentrated focus juxtaposed against voids and absences. By looking so intently at Bartram's drawing practice and how it informed his view of nature, I admittedly give less attention to other important aspects of his work.

Bartram's understanding of nature, for instance, was shaped in part by his Quakerism. John had been read out of the Darby Meeting in 1758 for questioning the divinity of Christ, and Kerry Walters has pointed out that the Bartram family's Quakerism was hardly doctrinaire. Still, both father and son retained a Quaker belief in an experiential relationship and knowledge of God, a relationship that could be made manifest in and through the natural world. Moreover, the Bartrams' most important patrons—Peter Collinson, John Fothergill, Robert Barclay—were noted members of the Society of Friends. Quakerism provided both a spiritual impetus and a social context for Bartram's study of nature, a topic that has been investigated by Amy Meyers, William Cahill, Larry Clarke, and Kerry Walters, but which goes unexamined here.[29]

Likewise, Bartram's knowledge of and writings on the Cherokee, Creek, Seminole, and other Native communities he encountered have long served as an important historical reference in anthropology and ethnography. Perhaps even more important is the role that American Indian guides played in his research. As he traveled through the Southeast, he often relied on the assistance of American Indians, gaining permission to botanize on their territories, relying on them for direction through the local topography, and gaining information on the properties and uses of different plants and animals. His natural history would not have been possible without their aid, and his *Travels* is struck through with an openness toward and admiration for the communities he encountered. Key figures such as Ahaya—the Oconee chief who gave Bartram the name "Flower Hunter" and allowed him to botanize in the Alachua Savanna—are given voice in the *Travels*, and the book's frontispiece of the Seminole warrior Mico-Chlucco is perhaps his only attempt at a human portrait. And yet, because this book focuses on Bartram's visual rhetoric and descriptions of nature, his connections to Native communities are addressed only tangentially. Kathryn H. Braund's and Gregory Waselkov's coedited volume, *William Bartram on the Southeastern Indians*, is an essential resource in this regard.

Also absent in this work is a deep investigation of Bartram's relationships with free and enslaved people of African descent. This area has been less studied than Bartram's engagement with American Indians of the Southeast, though it has received the attention of Sharece Blakney, Joel Fry, Kerry Walters, Christopher Iannini, and Monique Allewaert, among others.[30] Bartram surely knew the family of George Hilton, a free Black gardener working at William Hamilton's nearby estate, The Woodlands, though there is no mention of them in Bartram's extant letters. Bartram's relationship with enslaved Blacks is more clearly documented. When he attempted to establish his plantation in East Florida in 1766, his father purchased Jack, Siby, Jacob, Sam, Flora, and Flora's unnamed son to labor there.[31] Bartram was also given title to a woman named Jenny in 1772 by his cousin Mary Bartram Robeson (daughter of Colonel Bartram) and her husband, Thomas Robeson. It appears Jenny was brought from North Carolina to Pennsylvania, but what her role was while there is not clear: she may have worked at the Garden or been sent out to labor elsewhere. We know only that Bartram asked his brother-in-law to sell her on his behalf when he set off on his botanizing journey for Fothergill.[32]

Just as Bartram was aided by Native communities in his southern travels, so too was he assisted by enslaved people laboring on local plantations, who were often "lent" to him as he made his way through difficult terrain. Unlike the American Indians he encountered, Bartram gave them no voice in his text, except when he described them as praising "the virtues and beneficence of their master in songs of their own composition."[33] Despite his framing of slavery as quite literally harmonious, we also know that Bartram was patently aware of the institution's deep cruelty and degradation, as both an observer of the practice and as a slaveholder himself. The nearly four years between 1761 and 1765 he spent with his uncle in North Carolina would have been especially instructive. Twenty-nine enslaved people, of whom at least eleven were children, were enumerated in Colonel Bartram's probate inventory—as well as "a percil of Slaves the number Not known," "1 Mouth piece to put on Negros," and "1 pair Iron hoppels for Negros."[34] Slavery's brutality, apparent even in this dry list, did not keep Bartram from relying on forced labor or from using humans such as Jenny as capital.

In the second half of his life, however, we see Bartram struggle to come to terms with his participation in slavery. By 1783 he had drafted an antislavery treatise, expostulating at length about the inhumanity of the institution. He accused "those cruel Beings who purchase the life & person [of] their fellow creature"—of whom he was one—of being "guilty of the highest ingratitude," since they depended on the forced labor of others for their wealth and well-being. His treatise concluded with a dire warning about this practice: "sooner or later, ye must render full retribution [for this sin], Whether Voluntarily by your own free Act, Or by a Mighty hand."[35] If the 1783 treatise suggests an attempt to make himself right with God, a letter from 1788 speaks to Bartram's desire to reconcile with those he had wronged. He wrote to Mary Bartram Robeson asking her to "please present my Regard to all the Families of the *Black People*; They were kind and very serviceable to me; I wish it were in my power to Reward their fidility & benevolence to me. I often Remember them; These acknowledgement[s] at least are due from me to them," suggesting both deep shame and earnest apology.[36] The progression of Bartram's views on slavery seems to align with his increasing recognition of the interconnectedness of all life. If his understanding of the natural world disrupted or dissolved boundaries between species, genera, and even kingdoms, how much more profound its effects on his understanding of humanity?

Although I cannot address every important facet of Bartram's long and productive life and career in this book, I do endeavor to plumb a particular notion of figuring in eighteenth-century colonial science, thereby filling an important gap in art history, the history of science, and American studies. In taking seriously the peculiarities of Bartram's graphic enterprise, I attempt to tease out his theory of visual representation and consider its formative role in his engagement with and understanding of nature. I rely on close attention to the images themselves, pressing on their points of visual and cognitive rupture, which mark the myriad influences that shape his project: his empirical experience of nature, his integration into eighteenth-century epistolary culture, and his intellectual and social milieu. While signaling his influences through such fissures, Bartram also adapted these sources to his own process of coming to know—and, indeed, of figuring—the natural world. Ultimately, in this book I present Bartram's drawings as palimpsests of both the corporeal and the conceptual, spaces in which he consciously interleaved embodied and abstract knowledge, world and representation. By using such an approach I hope fundamentally to alter our understanding of his drawings and their strangeness, while also highlighting the contested terrain of scientific visualization in the eighteenth century.

Chapter 1

The Asymmetry of Transatlantic Natural History

"Our americans have very little tast for these amusements I cant find one that will bear the fatigues to accompany me on my peregrinations," John Bartram wrote to the Reverend Alexander Catcott in May 1742.[1] So much about transatlantic science in the eighteenth century is encoded in this lament from William Bartram's father. It points to the abiding European interest in the natural resources of Britain's American colonies, while underscoring how colonists often failed to match that interest. This incompatibility derived from the lack of a well-established scholarly infrastructure in British North America, so distant from the celebrated scientific societies, natural history collections, and printing houses in Europe. At the same time, this Old World infrastructure was reliant on New World collectors, those who had access to a natural world that most European virtuosi would never encounter directly, except through seeds, letters, drawings, and specimens shipped from abroad.

John Bartram's note to Catcott also highlights the difficulties settler colonists faced, both physical and financial, in procuring specimens from a far-flung wilderness. The ability to collect flora and fauna required local connections to like-minded naturalists who could serve as interlocutors and travel companions. William Bartram—who had just turned three years old when John Bartram wrote to Catcott—was raised to be just such an interlocutor and traveling companion, one possessing both the curiosity about the natural world and the physical stamina to accompany his father on trips to New York, New England, and into Britain's southern American colonies. It was through John's guidance and training that the younger Bartram was introduced to the study of nature and to the network of amateur and professional naturalists who would underwrite that study. Yet the asymmetry of access between colonists and European virtuosi in the eighteenth century also played a formative role in William's education.[2] In this chapter I examine the continual negotiations of that uneven and occasionally uneasy relationship between the Bartrams and their European peers.

Kind Frd John Bartram

That John Bartram himself was part of this vast Euro-American network seems on its face unlikely. He was born to Quakers in rural Darby, Pennsylvania, in 1699, where he received an indifferent education. He was orphaned by age twelve and was raised primarily by his paternal grandmother; upon reaching his majority, he married and set up as a farmer on land he had inherited. But after only three short years of marriage, his wife died, as did their first son, and by 1727 he was left widowed with a toddler. John decided to rent out the Darby property and relocated to farmland in Kingsessing, just across the Schuylkill River from Philadelphia. He remarried in 1729 to Ann Mendenhall, with whom he would have nine more children, William among them.[3] Although these early years gave little indication of John's future contributions to natural history, he would ultimately become, through a confluence of social connections and fortuitous timing, a key figure in the transatlantic trade of plants.

Perhaps most essential to John's development as a naturalist were his contacts with merchants in Philadelphia's city center, including Joseph Breintnall, the secretary of Philadelphia's Library Company. Founded in 1731 as an outgrowth of Benjamin Franklin's Junto, the company was the first membership library in British North America and counted among its members some of Philadelphia's most eminent citizens. The Library Company was an early step toward re-creating the scholarly infrastructure that existed in Britain by amassing a collection of textual and material resources for members' study and discussion. As its secretary, Breintnall worked directly with the London-based textile merchant Peter Collinson, who served as the Library Company's agent abroad, procuring books and other resources for its collection.

Although Breintnall and Collinson obviously shared interests in trade and library building, they were also connected by their fascination with American natural productions, many of which were still unknown in Britain. Breintnall had experimented with making impressions of plant leaves by inking them and running them through a press, but Collinson wished for more. He asked Breintnall whether he knew of someone who might be able to collect plants, seeds, and specimens of American nature, both for himself and for his fellow plant enthusiasts in Europe. John Bartram, who had already begun to develop a reputation for his in-depth knowledge of local flora, became that person.[4]

As a merchant, Collinson was impressively well connected, and his vast correspondence regarding trade also provided a vehicle for his inquiries about natural knowledge. He often requested that his faraway connections send specimens and seeds of exotic plants, which he propagated first in his garden at Peckham, south of London, and later at Mill Hill, his property to the north of the city. He also shared them with his colleagues, other Fellows of London's Royal Society for Promoting Natural Knowledge, Britain's premier scientific institution in the eighteenth century. John Bartram's epistolary introduction to Collinson would connect him to important planters and collectors in the colonies and to some of the most celebrated authors, artists, and naturalists in Europe. The president of the Royal Society Sir Hans Sloane, the Dutch naturalist Jan Frederik Gronovius, and the era's preeminent systematist, Carl Linnaeus, would all become important correspondents. So, too, would the English artist naturalist Mark Catesby, whose years traveling in the North American colonies formed the basis for his *Natural History of Carolina, Florida, and the Bahama Islands* (1731–1743), and George Edwards, the beadle for the Royal College of Physicians and the author of *A Natural History of Uncommon Birds* (1743–1751).

The earliest letters between Peter Collinson and John Bartram have been lost, but those that remain reveal how Collinson widened not only Bartram's social sphere but also his intellectual sphere. Collinson sent writings on natural history as well as seeds and plants from Europe, and he encouraged his American friend to make more and more extensive botanizing journeys, on which Bartram gained firsthand experience of different ecosystems in North America. In the mid-1730s Bartram began exploring beyond the outskirts of Philadelphia, traveling southeast to the cedar swamps of New Jersey and northwest along the Schuylkill River. Late in September 1738 he journeyed through Maryland to Virginia, staying with connections of Collinson's. In advance of John's trip, Collinson assured him that he had "sent thee Circular Letters to all my friends" and instructed him to observe some geological particulars and to collect information on the umbrella tree and ginseng.[5]

After his return in late October 1738, Bartram provided Collinson a synopsis of his trip along with a host of collected materials, including a mixed packet of perennial seeds, "turtle eggs & several roots packed up . . . [and] a box of insects & jar of papaw flowers & fruite." Clearly he hoped this offering would be sufficient to encourage additional support from Collinson for future botanizing. "If you pleas to order me to new england next fall," he wrote, "I am not much against it having health & prosperity allso I should be glad to have letters of recommendation to thy friends there." He noted in this letter his willingness to provide plants to Gronovius and Linnaeus as well, as he "perceive[d] thay are curious & ingenious botanists."[6] His travels—often paved by Collinson's introductions—would ultimately take him from Onondaga, New York, to the Florida peninsula, where he collected observations and natural productions for assessment by the curious in London, Leiden, and Stockholm.[7]

But as he mentioned to Alexander Catcott, these trips were arduous—physically demanding and often isolating. His 1738 trip through Maryland and Virginia covered over a thousand miles, most of it traversed alone. Matters had changed by 1754, and John explained in a letter to Gronovius how his son William was now a regular traveling partner: "I have a little Son about fifteen years of age that has traveled with me now three years & readily knows most of ye plants that grows in our four governments."[8] William was John's fourth son to survive past infancy, yet his older brothers seem not to have joined their father on long trips to collect natural history specimens. William was clearly his father's acolyte. As early as 1753 John and William traveled to Coldengham, the home of New York's colonial governor, on their first extended botanizing excursion. In covering the more than 150 miles from Kingsessing, Pennsylvania, to Fort Montgomery, New York, the Bartrams together amassed such an impressive parcel of seeds for Collinson that he had the accompanying itemized list published in the *Gentleman's Magazine*. The journal celebrated "the largest Collection that has ever been imported into this Kingdom," and the published list detailed receipt of one hundred different seed types.[9]

The early 1750s also marked William's first natural history drawings for Collinson, who commended him for "his pretty performances."[10] These early drawings typically featured local birds, as well different types of American trees.[11] Collinson seemed eager to welcome William into his extended circle of naturalists, just as he had previously introduced John's third surviving son, Moses, into his merchant network. (Moses traveled to London in 1751 with the intent of becoming a merchant seaman and was therefore the only Bartram ever to meet Collinson in person.) Collinson promised to show William's drawings to his friends and colleagues—including George Edwards, who had just published the fourth volume of his *Natural History of Uncommon Birds*—for perusal and assessment. Collinson further encouraged William's work by sending fine paper and drawing materials.[12]

Fragments and Particulars

Even as John Bartram and Peter Collinson became close, if geographically distanced, friends, their relationship remained inherently asymmetrical. The Bartrams provided keenly sought-after specimens of an unfamiliar and otherwise inaccessible natural world, and in turn Collinson offered them access to resources and connections to scholarly institutions that they could not find in the colonies. When John indicated that he wished to start an organization in Philadelphia on par with the Royal Society in London, Collinson attempted to dissuade him. Such a society would require the ability, Collinson noted, "to draw Learned strangers to you to teach Sciences . . . which can't att present be complied with Considering the Infancy of your Colony."[13] The American Philosophical Society was eventually established in Philadelphia in 1743, yet European virtuosi continued to view their American correspondents as rustics, incapable of making sense of the wilderness around them. According to them, American nature could be rendered intelligible only in Europe.

As Parrish observes, the asymmetry of transatlantic natural history profoundly shaped the relationship between Collison and the Bartram family.[14] The early Collinson–Bartram correspondence especially reveals how Collinson's curiosity about North America was shot through with a condescending dismissiveness. "Is there any account of the panthers[?]" he eagerly asked John in a letter from December 1737. "Do they attack man or cattle to see a live one I presume is not very Common to Europeans." Yet in the letter's subsequent paragraph his inquisitiveness disappears, as he adopted a patronizing air: "I shall first take notice of thy Request to buy Tournefort I have inquired and there are so many books or parts done as come to 50 shillings. . . . I shall be so friendly to tell thee this is too much to Lay out besides now thee has got parkinson & Millar, I would not have thee puzle with others for they contain the Ancient & Modern knowledge of Botany. Remember Solomons advice, in Reading of Books there is no End."[15]

Collinson referred here to John Parkinson's *Theatrum botanicum* (1640) and the first edition of Philip Miller's *Gardeners Dictionary* (1731), which he and Pennsylvania statesman John Logan had given John Bartram as gifts.[16] These were practical volumes, the former delineating the histories and properties of different plants and the latter providing instruction on their cultivation. The text John requested, Joseph Pitton de Tournefort's *Elemens de botanique* (1694), is a theoretical one that proposes a taxonomy of plants based on flower morphology. Although there may have been other reasons Collinson was reluctant to procure it (in the 1730s the English generally preferred the plant taxonomy of their countryman John Ray over Tournefort), his counsel against too much reading suggests that at first he saw John solely as a specimen collector rather than a fellow man of learning.[17] In a letter from January 1735, Collinson instructed John to send duplicates of dried specimens, so that he may "get them nam'd by our most knowing Botanists & then Returne them again," which, he maintained, "will Improve thee more then Books."[18] Threading through their correspondence is a generally held belief that the American colonial was unable to propose theories or patterns of natural organization. He need only send undigested observations and specimens abroad, where they could be transformed into natural knowledge and returned to America for educative purposes.

In his own letters, John's tone often slides between deferential and defensive. He adopted at times a self-effacing rhetoric, referring to his remarks as "rambling observations" and "dross" in need of organization and purification.[19] At other times, he groused about Collinson's incessant requests for new species, observing that he could not simply invent them. In another letter he lectured Collinson on the effort involved in collecting seeds and specimens: "Thee supposed me to spend 5 or 6 weeks in colections for you . . . but I assure thee I spend more than twice that time anualy. . . . you are not sensible of the fourth part of ye pains I take to oblige you."[20] All the same, John knew that his observations would need a social and institutional imprimatur in order for them to be accepted. Collinson, as a Fellow of the Royal Society, could provide that legitimation.

Indeed, Collinson did play a key role in bringing John's observations into the public sphere, as with the journal Bartram had written of a botanizing tour taken in summer of 1743 to Onandaga, New York, then up to the English trading post of Oswego. Although the trip was ostensibly a diplomatic mission to the Five Nations (the Mohawk, Oneida, Onondaga, Cayuga, and Seneca, now known as the Haudenosaunee), Bartram's primary aim was to consider the natural productions of the region and to assess it for the Pennsylvania colony's westward expansion. He sent his journal to Collinson in 1744, but the ship was captured by the French, which forced him to rewrite and resend his observations.[21]

It was only in 1751 that the manuscript finally made it into print, under the title *Observations on the Inhabitants, Climate, Soil, Rivers, Productions, Animals, and Other Matters Worthy of Notice, Made by Mr. John Bartram, in his Travels from Pensilvania to Onondago, Oswego, and the Lake Ontario, in Canada.* John's triumph of seeing his observations published may have been somewhat mitigated by the book's condescending preface, written by John Fothergill, the physician and Royal Society Fellow who would become William Bartram's primary patron beginning in the late 1760s. "It is a misfortune to the publick, that this ingenious person had not a literal education, it is no wonder therefore, that his stile is not so clear as we could wish." Still, Fothergill observed that there was some evidence of "good sense, penetration, and sincerity" that made the publication worthwhile.[22] This framing of John Bartram as an unpolished farmer, in need of explanation or editing, was not uncommon.

Collinson regularly brought John's work to his friends and colleagues at the Royal Society, and it was sometimes featured in the society's publication, the *Philosophical Transactions*. Letters on the fangs of rattlesnakes and various freshwater and saltwater mussels were published without commentary, but in other instances Collinson simply summarized John's observations, drawing out details from their ongoing correspondence and fashioning them into a narrative that may or may not give credit to his source.[23] For example, in a 1764 article on the cicada of North America, Collinson presented the observations as though they derived from his own empirical experience. "It is surprising to see how quick [the female cicadas] penetrate into hard wood, and croud it full of eggs," he wrote, as if the surprise on seeing this procedure was his own. Further on in the article he noted: "How she deposites the eggs . . . was difficult to discover, . . . but my ingenious friend John Bartram, observing her, in the beginning of this operation, took a strong woody stalk of a plant, and, presenting it to her, she directly fell to work upon it. . . . It was very wonderful to see how dexterously she worked her dart into the stalk, at every puncture dropping an egg."[24] The description reads as though John's only contribution was to ascertain how the female cicada laid her eggs, not that he had provided the bulk of the information that Collinson presented.

More commonly, however, John's information was neither presented independently in the pages of the *Philosophical Transactions* nor wholly appropriated; rather, Collinson would preface the letters with some reinforcing phrases, such as "Mr. *John Bartram*, a diligent Observer of natural Productions," to highlight the American colonial's reliability and accuracy.[25] Collinson would then take his observations and generalize from them, drawing in other resources and published texts to highlight his own capacity for intellectual synthesis.

In particular, a pair of papers on the European and American "libellae" (likely dragonflies or damselflies) reveals the extent to which Collinson framed and legitimized John's observations. In the first of these papers, read before the Royal Society in February 1750, Collinson proffered his own study of the life cycle of the European insect: "About the Beginning of May, I observed many deformed Water-Insects . . . creep up out of the water." The nymph's "back splits open; and from this deformed Case creeps out a beautiful Fly, with shining transparent Wings." Although Collinson's essay begins with a personal observation (his description of a specific instance of molting nymphs), it continues in a more generalizing vein. Collinson presented not the actions or behaviors of a specifically observed hatch of dragonflies but those of the whole of the species, revealing his capacity to synthesize empirical experience into abstract knowledge. At the conclusion of his report, he affirmed its accuracy by recourse to a well-respected entomological text, René-Antoine Ferchault de Réaumur's *Mémoires pour servir à l'histoire des insectes* (1734–1742), highlighting the accord between his own observations and those "curiously described and delineated by that excellent Naturalist Mr. *Reaumur*."[26]

If the first paper illustrates Collinson's ability to generalize natural knowledge from his personal observations, the second paper reveals how he also did so from John Bartram's. Presented before the Royal Society in April 1750, this paper provides Bartram's remarks on the libellae in North America, yet it does not begin as a firsthand account. Collinson introduced his correspondent's observations with a short synopsis that foregrounds the essential information he believed they offered—namely, that the American and European insects are the same, except the American species spends more of its life as nymphs. "By the Specimens before you, the May-flies of *America* have no very remarkable Difference from ours," Collinson noted, "excepting a few Days in the Fly State, they live all the Year a Water Insect." Following Collinson's introduction, the essay proceeds in journal format, beginning with John Bartram's records from when he observed that the nymphs had first attained wings to the laying and hatching of their eggs. Bartram's particular remarks are generalized and transformed into natural knowledge for the Royal Society by Collinson's introduction, without which they would merely be a sequence of firsthand observations. As Parrish points out, Collinson's direction to the reader to consider "the Specimens before you" underscores their fragmentary nature. Although the phrase probably refers to actual material specimens presented at the Royal Society meeting, it also directs the reader to Bartram's remarks, so akin to a collection of insect specimens.[27]

Collinson provided a similar service of legitimation for William. By January 1758 he had a drawing of William's engraved and printed in the *Gentleman's Magazine*, contextualized with Collinson's own verbal gloss (figs. 1.1 and 1.2).[28] His description vouches for the correctness of William's portrayal of the common musk turtle (*Sternotherus odoratus*), which had hitherto been unknown in England: "The tortoise represented . . . is a non descript animal, not to be met with in *Catesby*, or any other writer that I know of. Wherefore I procur'd these accurate drawings to be taken from a living subject, by *William Bartram*, an ingenious young man."[29] Collinson outlined the turtle's habitat, its color, the segmentation of its shell, and the sharp hornlike protrusion on the tip of its tail. From these characteristics, Collinson—the synthesizer of William's observations—took it upon himself to christen the species *Testudo Pennsylvanica cauda cornu armata*, or the Pennsylvania horn-tailed turtle, which in truth was merely his Latinizing of the name that the Bartrams had given the specimen. Collinson's Adamic act thus transformed William's rendering of a nondescript into an identifiable species that could be understood and categorized.[30]

George Edwards, with whom Collinson had shared some of William's drawings, would also shepherd the young Pennsylvanian's observations into the public sphere. Edwards had begun to receive his own shipments of specimens, drawings, and descriptions from the younger Bartram in June 1756 and began incorporating them into his three-volume *Gleanings of Natural History*

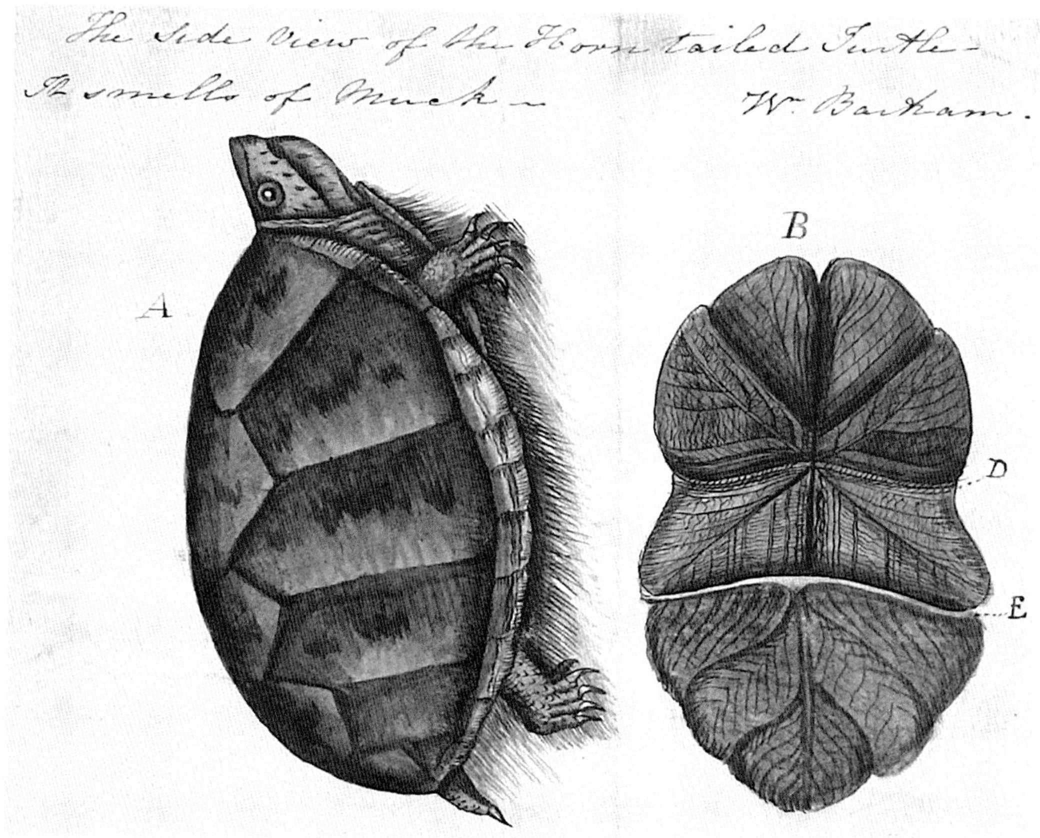

FIG. 1.1. William Bartram, *The Side View of the Horn Tailed Turtle* [*Sternotherus odoratus* or common musk turtle], 1755. Courtesy of Arader Galleries, New York.

(1758–1764)—a continuation of his earlier *Natural History of Uncommon Birds*. In the first volume of *Gleanings*, William's materials only confirmed information Edwards had gathered elsewhere; in the second and third volumes, however, they served as the basis for a number of the illustrations and textual descriptions. Plate 300 of the book, which depicts the black-throated green flycatcher and the black-and-white creeper, were drawn directly from specimens Bartram had provided: "These two rare and beautiful birds (which I believe to be non-descripts) were altogether unknown to me, till I had the pleasure of receiving them from Mr. Bartram." Edwards's detailing of the birds' diet and migration patterns all derive from his young colleague's firsthand observations.[31]

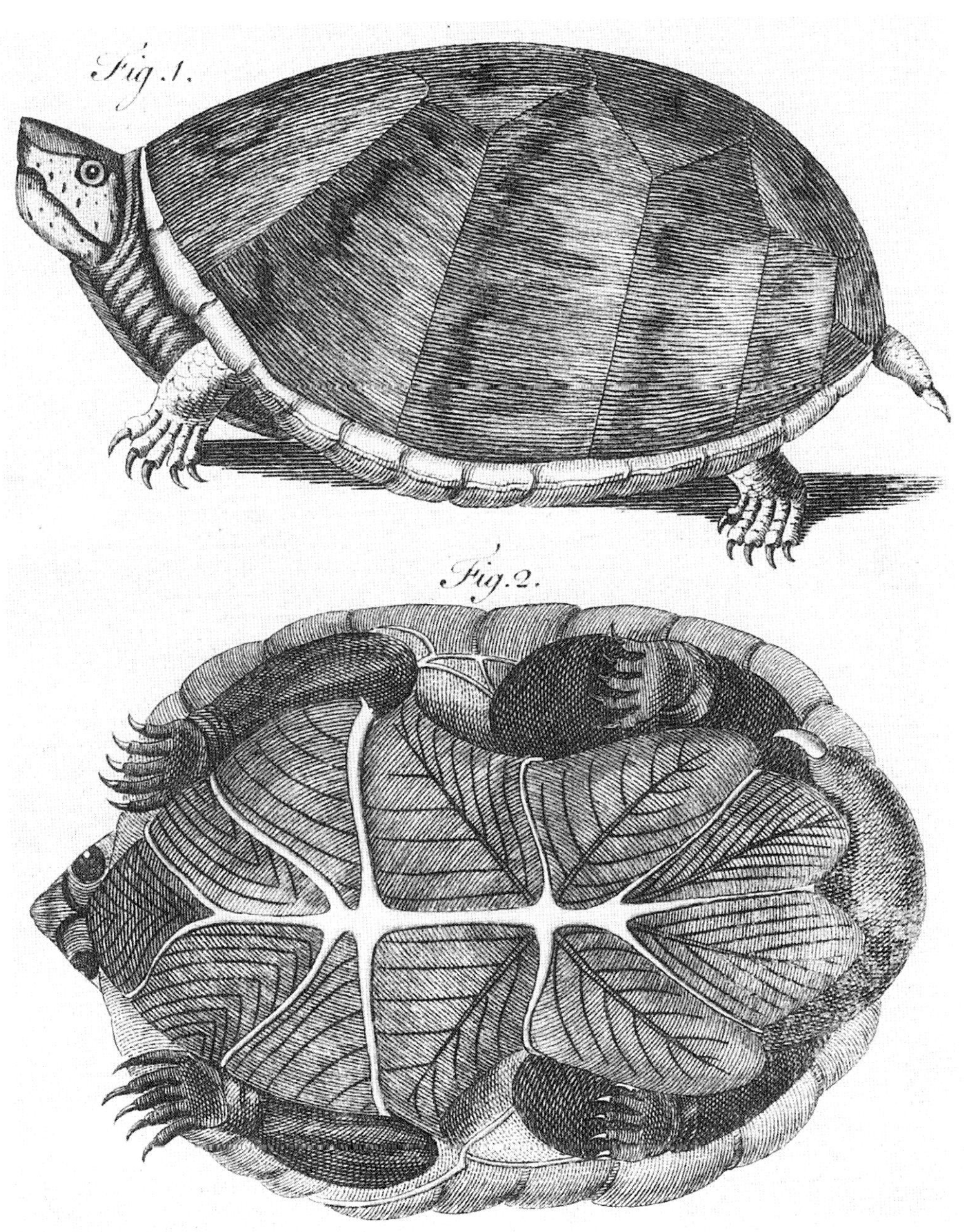

FIG. 1.2. *Testudo Pennsylvanica.* From *Gentleman's Magazine* (Jan. 1758). General Research Division, New York Public Library.

Occasionally William Bartram was unable to secure specimens of the various birds he wished to bring to Edwards's attention. In such instances, he provided drawings of the species for his mentor's assessment, which likewise found their way into *Gleanings*. Plate 304, for instance, depicts the white-throated sparrow, "figured of the size of life." Yet in this figuring Edwards is at one remove, since the etching "is taken from a neat drawing in colours, done by Mr. William Bartram of Philadelphia in Pensilvania." Edwards continued, "I have never seen the bird; but thought Mr. Bartram's drawing of it very curious, and have reason to be satisfied as to his veracity and accuracy."[32] Edwards vouched for Bartram to the reading public, indicating that his work should be accepted as both truthful and correct.

But how could Edwards be so certain on these points? In the text accompanying plate 291 of *Gleanings*, which depicts a female marsh hawk (*Circus cyaneus* or northern harrier; fig. 1.3), Edwards noted, "I have not seen the Bird itself," and that he again had to rely on a rendering provided by Bartram. He assured readers, however, that he had "great reason to believe Mr. Bartram very correct in his drawing, and exact in his colouring, having compared many of his drawings with the natural subjects, and found a very good agreement between them."[33] Also featured in the plate is a female rice bird (*Dolichonyx oryzivorus* or bobolink) held fast by the marsh hawk's talons. Edwards claimed, "the Rice-Birds are thrown into the plate by way of decoration; they are copied in miniature from Catesby's book [fig. 1.4], though in attitudes different from his figures."[34] Much like the case of the white-throated sparrow, Edwards's textual frame for the marsh hawk illustration reveals how he legitimized Bartram's observations. His text provided an assurance of Bartram's truthfulness and accuracy, as did the plate's incorporation of Catesby's recognizable rice birds, since these accepted and authoritative illustrations would already read as true.[35]

It is only by comparing Edwards's plate with the original drawing, however, that we see just how truthful and accurate Bartram was (plate 1). Edwards transformed Bartram's sleek huntress into a heavier bird, one that is broader in the breast and shorter in the tail and leg. In Edwards's illustration the bird's facial disc—the concave circle of feathers that collects soundwaves and helps it locate prey—extends down the neck, making it read as a field mark rather than an essential part of the marsh hawk's sensorium. Moreover, the ratio of primaries (outer flight feathers on the wing) to secondaries (inner flight feathers) is entirely different in Edwards, with fewer

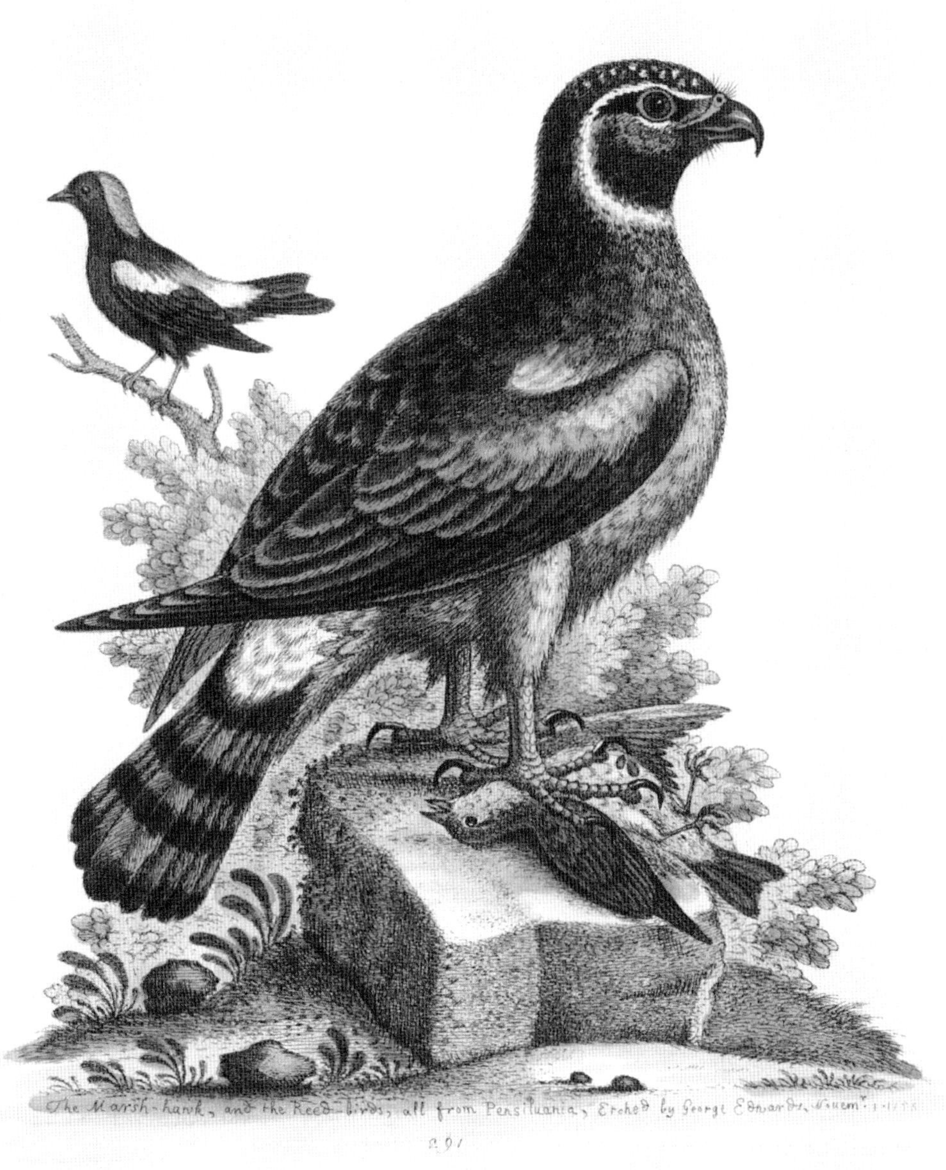

FIG. 1.3. George Edwards, *Plate 291: The Marsh-Hawk, and the Reed-Birds*. From *Gleanings of Natural History*, vol. 2 (London, 1760). Digital Library for the Decorative Arts and Material Culture, University of Wisconsin–Madison.

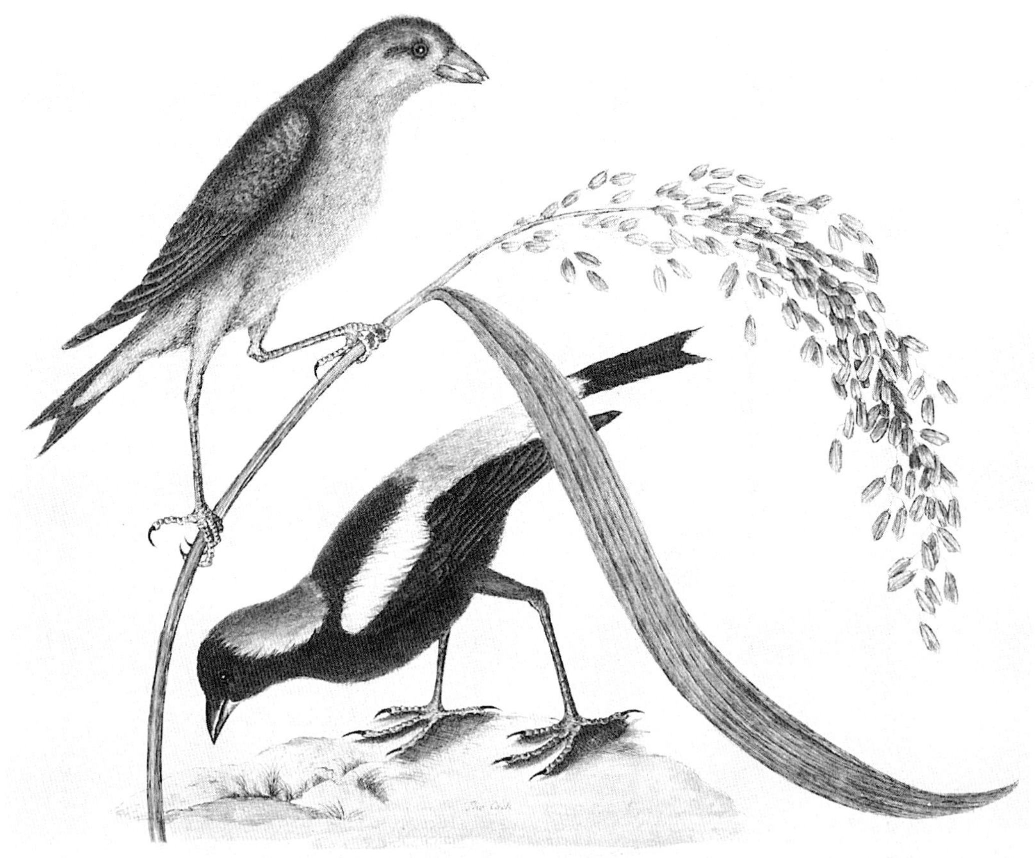

Fɪɢ. 1.4. Mark Catesby, *Plate 14: Hortulanus Caroliniensis: The Rice-Bird*. From *Natural History of Carolina, Florida, and the Bahama Islands*, 2nd edition, vol. 1 (London, 1754). Digital Library for the Decorative Arts and Material Culture, University of Wisconsin–Madison.

and wider feathers than the bird possesses. Beyond these critical errors, Edwards conveniently obscured the fact that it was Bartram who inserted the rice birds in the original composition. By making it seem as though he himself had "thrown" them in "by way of decoration," Edwards took credit for this visual gesture of accuracy, as though his protégé's observations needed not just textual but also graphic intervention in order to become knowledge. Edwards's interventions were clearly neither necessary nor knowledgeable: he significantly altered Bartram's portrayal of the marsh hawk based on no discernable evidence, unintentionally misrepresenting the bird in the process.[36]

There is no extant response from the Bartrams regarding the misrepresentation of the marsh hawk in *Gleanings*. Their silence on this point may have derived from a recognition that John and William were deeply dependent on the patronage of figures such as Collinson and Edwards and had to tread thoughtfully. Collinson underscored this point in 1765, when he helped John Bartram win a royal pension. John used this financial windfall to complete a botanizing trip through East Florida, a newly British colony following the Seven Years' War. In October of that year, he and William traveled along the St. Johns River in the Florida peninsula, taking notes on the region's soil and natural productions. Collinson waited anxiously for reports. Though the first box of materials was disappointingly lost, others provided Collinson "much Entertainment," as did John's written accounts.[37] Collinson surely imagined he would shepherd these observations into the public sphere, and when he discovered John's journal had been published without his knowledge, he was deeply upset. "If thou hadst an inclination to oblige a Friend that has Served thee & Family for so many years I should expect to have heard thee saye I know my Friend Peter would like to Peruse the Journal . . . but if this was to great a Favour the Least thou couldest have done was . . . to Send & Complement Mee with a Book."[38] Collinson warned that the pension he had helped John obtain could be rescinded, indicating that he was not simply some middleman who could be circumvented.

Transformations and Transmutations

John and William Bartram chafed at times against their colleagues' intellectual processing, and they both made clear, in their own ways, that studying nature from a distance necessarily compromised the endeavor. One could never experience the intransitive properties of place simply by amassing seeds, specimens, descriptions, and drawings. Writing in 1741 to Mark Catesby, John noted that the point of his botanizing expeditions—even to sites that were difficult to access—was to see nature in all her "transformations & transmutations."[39] The natural world was not static and immutable, he averred, but always in transition. As George Edwards's marsh hawk etching indicates, objects and observations sent from abroad could only ever be incomplete fragments, and ones that were susceptible to misinterpretation.

Still, Collinson, Edwards, and other London naturalists seemed at times to overlook the dilemma posed by these fragments, as though American nature could be transferred wholesale to Britain. In a letter from September 1741, Collinson noted that the trees from Britain's American colonies were so densely planted at Thorndon, the Essex estate of Robert James, Lord Petre, "one cannot well help thinking he is in North American thickets."[40] He followed up this observation by indicating he likewise felt in the midst of South America, so prolific were the pineapple, guava, and lime trees propagated in Thorndon's stoves. The rhetorical move Collinson makes—his quick, imaginative shifting from North to South America—implicitly recognizes that the flora on Lord Petre's estate can only give a sense of, yet never truly capture, the plants' points of origin.

In other instances, the Bartram–Collinson correspondence patently addresses the challenges of long-distance natural history, and the letters are replete with mention of living specimens killed in transit or transplanted roots and seeds that would not sprout. Regarding one particular shipment, Collinson noted with great disappointment how the "Curious Collection of living plants" John Bartram had sent was wholly devastated by rats that had nested among them, killing everything except for a single root.[41] Rats were great impediments to the safe shipment of botanicals, but mold and insects were a nuisance too. Rotting packaging could compromise the seeds, and insects could nibble away at leaves and roots or even, sometimes, the letters accompanying them.[42] Collinson noted, for instance, how one of John's letters was so riddled with holes due to "Mischievous insects" that he struggled to make out the text.[43]

Even when American seeds, plants, and letters arrived whole in London, that anything might come of them remained uncertain. Collinson often requested a list of plants he wished to propagate and then followed up with a letter describing his success or lack thereof. In the same letter that conveys his disappointment at the insect-eaten letter, he recounted to John, "Of the Seeds thou Sent the Rose Laurell are Come up and are very thriving, Red Cedar by thousands White Cedar a few—Sheeps Turds none—must send a young Tree."[44] If foreign plants did successfully establish themselves in distant locales, what might happen was still anyone's guess. In 1759, for example, John Bartram sent Collinson a disquisition on how non-native plants often turned into invasive weeds that spread from his garden and overtook his pastureland. Collinson read his tract with great interest. "See what Climate & Soil does," he observed, noting how species so often reacted differently in different environments, depending on soil and situation.[45] Just as English plants sometimes behaved strangely when set in American soil, American plants might also grow in seemingly aberrant ways when propagated abroad. When Bartram sent Collinson seeds of meadow-sweet, Collinson reported that he was able to grow the plant but that it would not bloom, and he inquired whether it flowered in the garden at Kingsessing.[46]

For American flora that could not successfully be propagated abroad or American fauna that could not safely be shipped alive, an alternative was to provide dry or wet specimens. The flower, leaf, and sometimes the fruit of plants could be pressed, dried, and enfolded in quires of paper for easy shipment. The shape would be altered, yet key characteristics of the flora would still be preserved through this process. Plants with larger fruits were more difficult to dry and press, so in these instances they were often sent as wet specimens. On Collinson's request, John Bartram provided fruit and flower of the pawpaw preserved in a jar of rum for Catesby in 1739. Collinson, however, also asked for detailed verbal descriptions to accompany the specimens, since the colors "fade so in paper, and change so in spirits" that they were nearly impossible to ascertain.[47]

Preserving animals for shipment was even more challenging. As Robert McCracken Peck explains, fleshy specimens were often preserved in spirits of some kind, much like the pawpaw packed in rum, but collectors tended to dislike the visual effect of this preparation. Dry specimens were preferred for larger vertebrates, from which the organs and skeletons were excised. The remaining skins could be dried in an oven and treated with some preservative—often tobacco, spices, or camphor—to protect them from insects. The skins were then mounted on a wire armature and stuffed with straw, cotton, or other fibers in an attempt to approximate the animals' original appearance. Peck notes that this process often distorted the organisms' forms, making them appear like "grotesque caricatures of their former selves."[48] George Edwards repeatedly noted throughout his multivolume *Natural History of Uncommon Birds* just how misleading such preparations could be. Specimens of the bird of paradise, he wrote, "are brought from the *East-Indies*, and are generally imperfect, which has caused them to be variously figur'd, and described as different Species."[49] The living characters of both flora and fauna were often massively altered through the preservation processes.

John Bartram noted as much in his correspondence with Collinson, indicating how a lack of experience with living biota, in situ, often led to a chain of misrepresentations. A key instance of this was Catesby's portrayal of the rattlesnake in his *Natural History of Carolina, Florida, and the Bahama Islands*. Rattlesnakes were a common topic of discussion for John Bartram and Collinson; Bartram provided observations and assessments of this exotic species, some of which Collinson published in the *Philosophical Transactions*. When John and William encountered a rattlesnake on their first extended botanizing trip to upstate New York in 1753, John, of course, relayed the experience to Collinson. The snake, John noted, was coiled in someone's hat, and he attempted to draw it out to its full length in order to see its body. The snake became angered and "drawed in his quoil again ratled & flattened his body by which his colors brightened much." John observed it was the same type of rattlesnake that Catesby had figured, yet the plate in his *Natural History* did not show the creature "in his greatest beauty." Although Catesby had surely seen living rattlesnakes while he was in North America, John Bartram assumed that he had used a dead specimen as his model. "Many laughs at ye story of changing color," but it "is realy a matter of fact."[50] (John was right that Catesby had not used a live specimen; his rattlesnake illustration is copied almost exactly from a watercolor by Nicolas Robert, which he would have seen in the collection of Hans Sloane.)[51] The self-assurance evinced by many European virtuosi in regard to American biota, Bartram lectured, was not always warranted.

The Grammar of Nature

The mutability of the natural world, and the ease with which specimens, descriptions, or even living plants might mislead, highlighted a key concern: how could any European-crafted taxonomy properly accommodate American flora and fauna? Would it not be more accurate to derive a taxonomic system based on close familiarity, in situ, of the biota one wished to describe? John Bartram's early, dismissed request to Collinson for a copy of Tournefort's *Elemens de botanique* suggests his interest in establishing his own botanical taxonomy, one rooted in his experience of North American nature.

Questions of classification were of great importance in the years that John Bartram and Peter Collinson corresponded. They would both be witnesses to and become participants in one of the most significant developments in the history of botanical taxonomy. In the 1730s, the systems most widely accepted were those established by Tournefort's *Elemens* and by John Ray's *Methodus plantarum nova* (1682). Which method one followed was often a matter of national identification. Naturalists in France and its colonies were more aligned with Tournefort, but throughout the British imperium naturalists tended to prefer Ray. The publication in 1735 of Carl Linnaeus's *Systema Naturae*, however, introduced a radical shift in the organization of the natural world. His system was not nationally delimited, and it would have profound effects on how biota was represented, both textually and visually, for the rest of the eighteenth century.

When Linnaeus published the first edition of *Systema Naturae*, he had just finished his training in botany and medicine at Uppsala in Sweden and had begun to work out his basic premises for categorizing natural productions. Although his larger divisions between the vegetable, animal, and mineral kingdoms were drawn directly from Aristotelian thought, those within each category were more innovative. With the vegetable kingdom Linnaeus, like Tournefort, turned his attention to the flower, but he focused specifically on the number and arrangement of its stamens and pistils (their male and female organs of reproduction) as the means for distributing plant life into classes and orders. Two years later, with the publication of *Genera Plantarum* (1737), Linnaeus organized all known plant genera within the outline articulated in *Systema Naturae*. Where plant class and order were determined by a flower's stamens and pistils, genus was ascertained by its calyx and corona, as well as by the plant's fruit and seed. Species within a genus could be differentiated by the form and position of the leaves, stems, and roots.

Although Linnaeus in his introduction to *Genera Plantarum* derided illustrations as pedagogically useless, he included in the book an image that translated his twenty-four plant classes into visual form (fig. 1.5). Created by Georg Dionysius Ehret, one of the more prolific and well-regarded botanical illustrators of the eighteenth century, the image (often referred to as the *Tabella*) reveals the system's simplicity. Notably, the rendered figures do not represent any known plant species but instead illustrate the abstract *idea* of Linnaean plant classes. The number, position, and proportion of the stamens are clearly delineated and—in some versions—originally highlighted with bright primary colors. Linnaeus conceded that the use of stamens as the basis for disposing plants into classes was a rational expedient rather than a natural division, and the tidy arrangement and arbitrary colors of Ehret's illustration indicate the system's conceptual, rather than natural, status.[52]

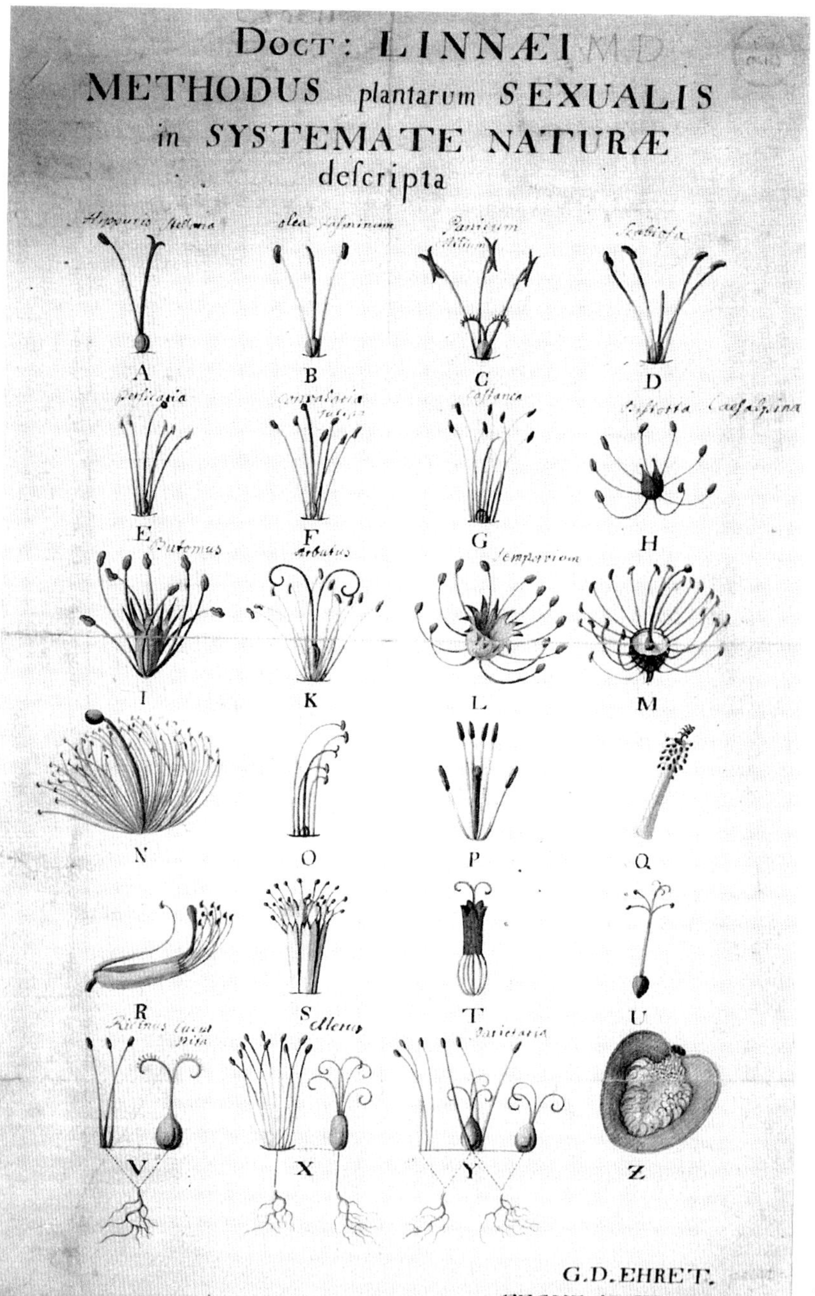

FIG. 1.5. Georg Dionysius Ehret, *Methodus Plantarum Sexualis in Sistemate Naturae Descripta*, 1736. © Natural History Museum, London.

The simplicity of Linnaeus's method, its reduction of plant identification to a limited set of "essential" characteristics, made it easily accessible to the novice as well as to the experienced botanist. Although some naturalists were reticent to adopt his system because of its perceived sexual impropriety, it ultimately became the primary plant taxonomy throughout much of Europe and its colonies by the mid-eighteenth century. Not surprisingly, its popularity exerted great influence on the visual representation of plants. If the identification of a species required a focus on certain characteristics and qualities and the elision of others, then illustrations, too, had to reflect this hierarchy of information. Questions of habitat, growth, and even color were viewed as being far less significant in the Linnaean system, which effectively untethered the plants from their places of origin and their lived character.

Both Peter Collinson and John Bartram were initially hesitant to adopt Linnaeus's method, but by the 1740s they had become correspondents in the Swedish naturalist's broad epistolary network. Still, Bartram observed that Linnaean taxonomy left something to be desired, complaining to Philip Miller that he found an "abundance of many plants that do not answer to any of [Linnaeus's] Genera."[53] In August 1753, he wrote to Collinson proposing to catalogue all North American forest trees according to his own system, baldly stating that neither ancient nor modern authors had made any "particular observation worth notice." In this letter, Bartram offered what he called a "specimen" of his method, using the hop hornbeam as an example. Unlike Linnaeus's more abstracted approach, in which flora is visually dissected and each of its essential characters independently described, he offered a narrative that tracks and describes the tree's shoots in the spring, the development of catkins (flower clusters) and fruit in the summer and fall, and the subsequent opening of the catkins the following spring. He concluded his narrative of the tree's developmental processes with a description of its form and dimensions at full growth, the character of its bark, and its general habitat.[54]

Bartram was greatly enthused about the prospect of accurately describing and categorizing North American forest trees, mentioning his endeavor to a number of European correspondents. Collinson, who received his "dissertation" on the topic, was not entirely convinced by his methodology. "The Discriptions are so exact and Natural that I am always delighted reading them but my Good Friend I must Impart to thee my doubts—I am afraid the Species are so multiplied that it will be a difficult task to distinguish them Here." He suggested that

Bartram was incorrectly dividing species based on habitat or stage of growth. If John were to "distinguish things aright," he would realize that there were far fewer species than the many his taxonomy had originally identified.[55]

John Bartram was not to be dissuaded. In a letter written at the end of 1754, he asked Collinson to submit to him any errors so that he might make corrections to his system. This letter was also accompanied with a peremptory "quire of specimens of our oaks with their acorns on in perfection," as well as a packet of drawings by William, depicting "most of our real species of oaks and all our real species of birches." The drawings were made "not according to grammar rules, or science but nature," John explained. They would reveal that he was not mistaken; that his system was wholly accurate. He pointed out that other artist naturalists, in wishing to follow preexisting conceits regarding the order of the natural world—whether Linnaean or otherwise—typically ended up making a hash of nature and depicting her productions "wildly unnaturaly." John leveled this claim specifically at Catesby and his representation of oak trees, which he argued did not clearly articulate the difference between summer and winter acorns (likely meaning species of oak that ripened acorns in one summer and those that required a winter dormancy before ripening in the second year).[56] He ultimately hoped that the Bartrams' combined efforts—that is, his arboreal taxonomy supplemented by William's drawings—might be formally reviewed and a publisher found for them.

The Bartrams waited in limbo for well over a year, and it was only in February 1756 that Collinson had any news to report: "But yesterday I had a Letter from Docr Gronovius He admires much the Drawings of the Oakes but He can get nobody to Engrave & print them. . . . Our Friend Ehret will do them he Tells Mee but I can't say when."[57] Such a promise was surely tantalizing, but neither the catalogue of American forest trees nor William's associated illustrations ever made it into print, and the whereabouts of much of this material is now unknown. As a result, it is impossible to know the full scope of John's taxonomic method or how William's drawing of oaks and birches illuminated it. John's system and the accompanying figures seem to have been viewed by Europeans largely as curiosities rather than as serious essays into structuring botanical knowledge.

Collinson's concern that John was not distinguishing characteristics properly—or, in fact, was distinguishing too many characteristics—indicates his uneasiness about the Bartrams' ability to rationalize and generalize. In this particular instance, he seemed disinclined to put out their work under his imprimatur, fearing it might invite derision from his fellow virtuosi in Europe. Despite the many natural fragments and observations that the Bartrams had provided to Collinson and others, their contributions could only serve as raw material that needed to be assessed by the knowledge-making apparatus of European naturalists and institutions, a process that inherently deemphasized or even effaced the Bartrams' original material and intellectual labor.

Although the descriptions and drawings of oaks and birches have not been located, there still exist several drawings of maples made by William in 1755–1756, which were likely part of the Bartrams' taxonomic project. Among these drawings is a depiction of the red maple (*Acer rubrum*; plate 2), which was first figured in 1691–1692 in Leonard Plukenet's *Phytographia*. Plukenet's uncolored engraving, depicting only the shape of the leaves, offers few of the tree's distinctive qualities, which has dazzling red flowers and seeds in the spring and early summer. Nearly sixty years after Plukenet, Catesby included an image of the red maple in volume 1 of his *Natural History*, featuring more, and more useful, detail. Here we see several of the flowers, as well as a cluster of winged seeds or "keys" (fig. 1.6). In 1755 the chief gardener at the Chelsea Physic Garden, Philip Miller, published his own illustration of the red maple, intending it as a corrective to Catesby's figure. Catesby, he claimed, represented "the Seed-vessells very perfect, but the Flowers are not very correct; the Stamina are stretched out too far from the Corolla, and [they] are ill-coloured."[58]

His sense of assurance notwithstanding, Miller's rendering is hardly an improvement (fig. 1.7). Not only do the leaves look more akin to silver than red maples, but red maples produce leaves only after the flowering stage has ended; Miller offered instead a temporal conjunction of a branch with both flowers and leaves. Catesby nodded to the various points of the tree's life cycle by depicting the flowers and seeds on different branches and by isolating the maple's leaf. Neither Catesby nor Miller correctly portrayed the red maple's blooms, however. The tree has separate male and female flowers, which grow on separate plants—or occasionally on the same plant but on different branches. Catesby used different branches only to distinguish stages of the tree's life cycle, not the different types of flowers it produces, and his illustration shows only the tree's male flower. Miller, too, depicted only one kind of flower, which oddly does not accord with either the male or the female blossom but seems to be of his own invention. In his drawing of the red maple, William Bartram depicted

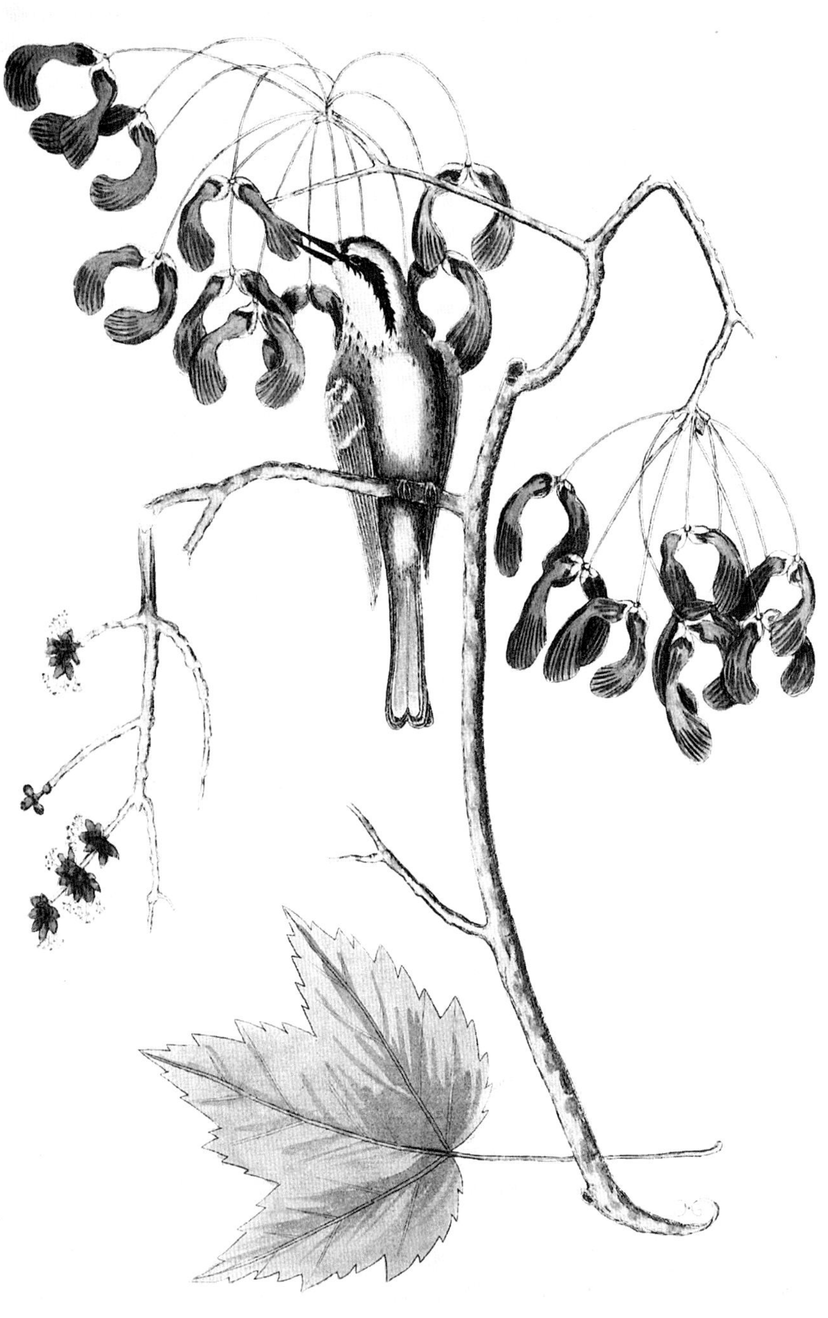

FIG. 1.6. Mark Catesby, *Plate 66: Yellow-Throated Creeper, The Red Flowering Maple*. From *Natural History of Carolina, Florida, and the Bahama Islands*, 2nd edition, vol. 1 (London, 1754). Digital Library for the Decorative Arts and Material Culture, University of Wisconsin–Madison.

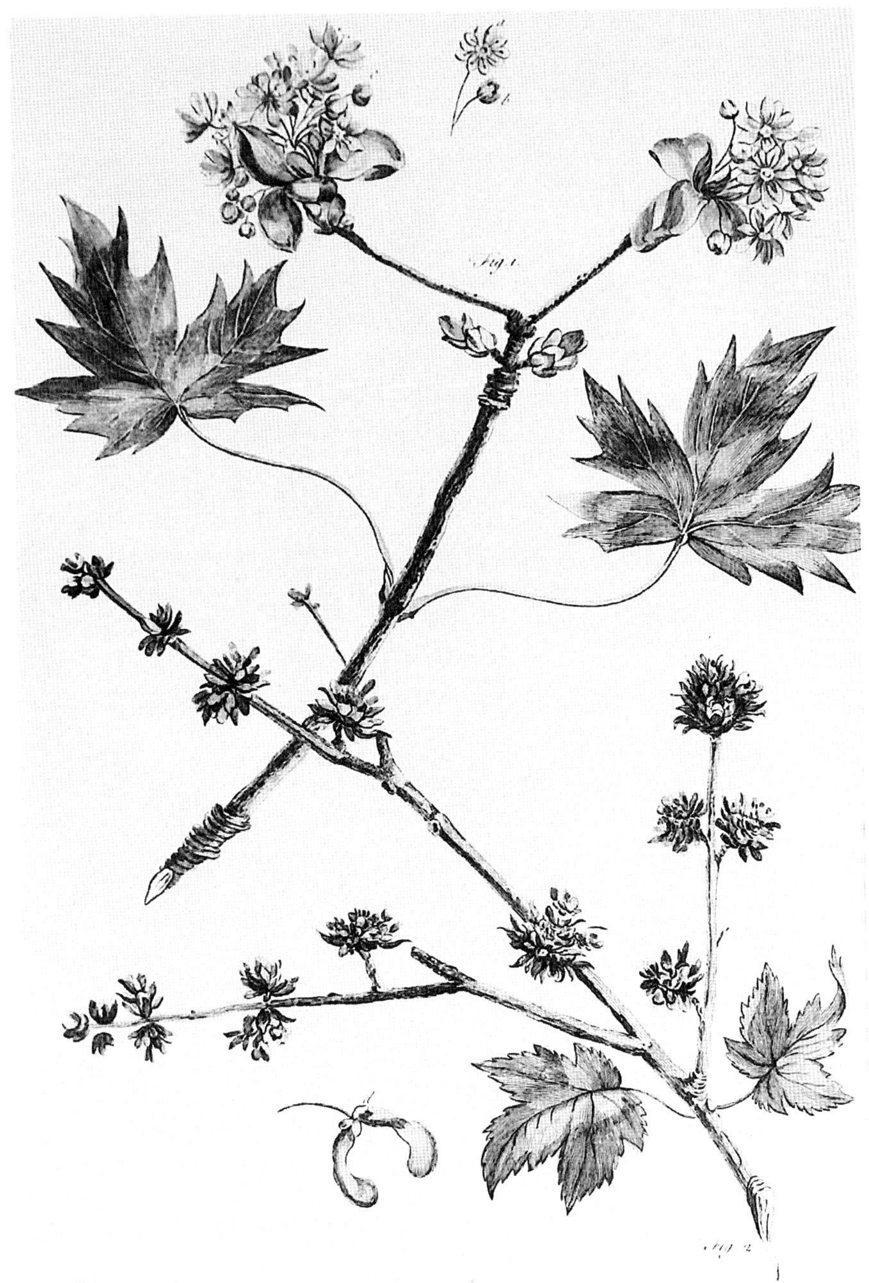

Fig. 1.7. Thomas Jeffreys, *Plate VIII.* From Philip Miller, *Figures of the Most Beautiful, Useful and Uncommon Plants,* vol. 1 (London, 1760). Courtesy of Arader Galleries, New York.

two branches, one for the tree's male flowers and the other for the female. The male blooms each possess a nimbus of stamens that suggests the plant's dynamic bursting forth from its winter dormancy, while the female flowers are pendant with pollen-collecting styles. Despite his youth—William was only seventeen when he made his drawing—he seemed to have ascertained the characteristics of the red maple with greater detail and accuracy than two of England's most celebrated botanists.

The void created by the missing arboreal taxonomy and most of the associated drawings makes it impossible for us to assess exactly how text and image related—that is, the ways in which John Bartram's taxonomy established the terms of the trees' visual representation. Judging by William's drawing of the red maple, we can surmise that the other images likewise conveyed the distinctive characteristics and living quality of the specimens, which John found so essential to his system of categorization. Although most of these works are no longer extant, William's drawings from the late 1760s onward indicate how he attempted to right the asymmetry of transatlantic knowledge, mapping out the means by which he came to know American nature. They also reveal his ability to craft representations of flora and fauna that were as dynamic and unexpected as the vibrant ecosystems he observed.

Chapter 2

William's Inimitable Picture

Shortly after attempts to find a publisher for the Bartrams' arboreal taxonomy came to nothing, William was apprenticed to the merchant James Child, an apprenticeship that left little time for drawing. His experience with Child seems to have been unremarkable. William worked hard and was well liked, and in 1761 he moved to Ashwood, North Carolina, to set himself up in business. John Bartram's half brother, Colonel William Bartram, had a rice plantation on the Cape Fear River, and William went to operate a plantation store there. Colonel Bartram's son, also William, came to Philadelphia to study materia medica. The exchange of sons may well have been John's suggestion: even though he had pushed William toward a career in trade rather than in botany, he surely would have been keen to have a reliable plant collector situated on Carolina's coastal plain.

As a merchant, William struggled to turn a profit. He anxiously tried to conceal his business troubles from his father, yet William's financial and emotional distress was common knowledge. When John began organizing his botanizing tour of East Florida in 1765, he invited William to accompany him, and William happily accepted. As the two Bartrams traced the length of the St. Johns River,

William became fascinated with the astonishing diversity of the East Florida landscape and decided to establish, with help from his father, an indigo and rice plantation along its banks. His ill-conceived foray as a slaveholding planter collapsed within a year.

The loss of the plantation allowed William once again to return his full focus to drawing and natural history.[1] Collinson, realizing that his young friend was at loose ends following his failed business endeavors, requested drawings that he thought might help him secure patrons for William. In July 1767 he specifically asked for a drawing of the American lotus (*Nelumbo lutea*), which he had been unable to establish in his garden at Mill Hill. He wished for a "picturesque figure," he wrote, that featured the plant's leaf, flower, and seed so that he might better ascertain its genus and species.[2] What Collinson meant by "picturesque" is not entirely clear from his letter; he may have wanted the flower depicted in a tableau or simply that he wanted a drawing that was as aesthetically pleasing as it was informative.

In response, William produced an elaborate pen-and-ink drawing of the lotus as it grows in its watery environs, capturing its development in the portrayed bud and blooms (plate 2).[3] To enhance the picturesque effect, he tucked beside it a vignette of a great blue heron and a Venus flytrap, each pursuing in its own way its respective quarry, with the bird dipping forward to catch a tiny fish and the flytrap's lobes open to entice a dragonfly. As delightful as the composition seems on its face, however, it is also deeply strange. The lotus blossoms tower overhead, dwarfing the great blue heron in the foreground, while the flat, floating lotus leaf tips forward, as though seen from above. Collinson was clearly charmed by the drawing—he claimed it "surpasses all" in a letter to John—but an undercurrent of perplexity also runs through his correspondence, suggesting that the work was unlike anything he had anticipated.[4]

Part of the drawing's strangeness is its inconsistent scale and its multiple points of view, which have led some scholars to interpret it as the work of an untrained eye and hand, of someone unfamiliar with period conventions of natural history representation.[5] William's status as an American colonial has sometimes encouraged such interpretations, yet his years of correspondence with British naturalists indicate that it was not an unfamiliarity with the visual conventions of natural history that he possessed but, rather, a looser, more ambivalent relationship to them. These conventions were premised on the principles of transparency and self-evidence, as though the natural world simply revealed itself to the innocent eye. Such visual rhetoric essentially minimized the efforts of colonials, who observed and collected American natural productions for study. William Bartram's drawing of the American lotus might be seen as a rebuttal to the era's received representational conventions, one in which he instead called attention to his own processes of exploration and discovery.

Nature's Self-Evidence

In eighteenth-century natural history, perhaps the most popular form of self-evident "representation" was no representation at all. As John Woodward articulated in his *Brief Instructions for Making Observations in all Parts of the World* (1696), travelers should "put [Samples of Plants] each *separately*, betwixt the leaves of some large *Book* . . . till they are sufficiently dryed, when a *weight* may be laid upon them to press or smooth them."[6] Collinson owned a copy of Woodward's book, and he provided almost verbatim direction in his earliest correspondence with the Bartram family.[7] For European virtuosi, pressed and dried plants were as important as verbal descriptions or visual renderings—they offered the specimen itself rather than the collector's impression or interpretation. Mark Catesby, for instance, followed the procedures outlined by Woodward during his American travels in 1722 (fig. 2.1). The lotus specimens he sent to the Oxford botanist William Sherard included minimal commentary, providing only the plant's references in the anonymous *Catalogus Plantarum Horti Medici Oxoniensis* (1648) and Leonard Plukenet's *Almagestum botanicum* (1696), as well as his name and the place and date of collection. Woodward's directions, which stated that one write only *"in what Country* the inclosed collection of Plants were gathered,"* indicate that the prepared specimen was intended to speak for itself.[8]

The botanical specimen—perhaps the most basic element in the construction of natural knowledge—also recalls a different form of supposedly natural representation: the nature print. Joseph Breintnall's experiments in inking and printing leaves and fabrics (fig. 2.2), which he had sent to Collinson in the 1730s, likely following the instructions in *The Art of Drawing, and Painting in Watercolours* (1731).[9] As the notations on a number of Breintnall's prints indicate, the intent was to capture the texture of the object as much as its contour. The indexicality of nature prints—that is, the direct physical relationship between the object and its representation—made them especially compelling to Benjamin Franklin, who developed a method of incorporating them into Pennsylvania's

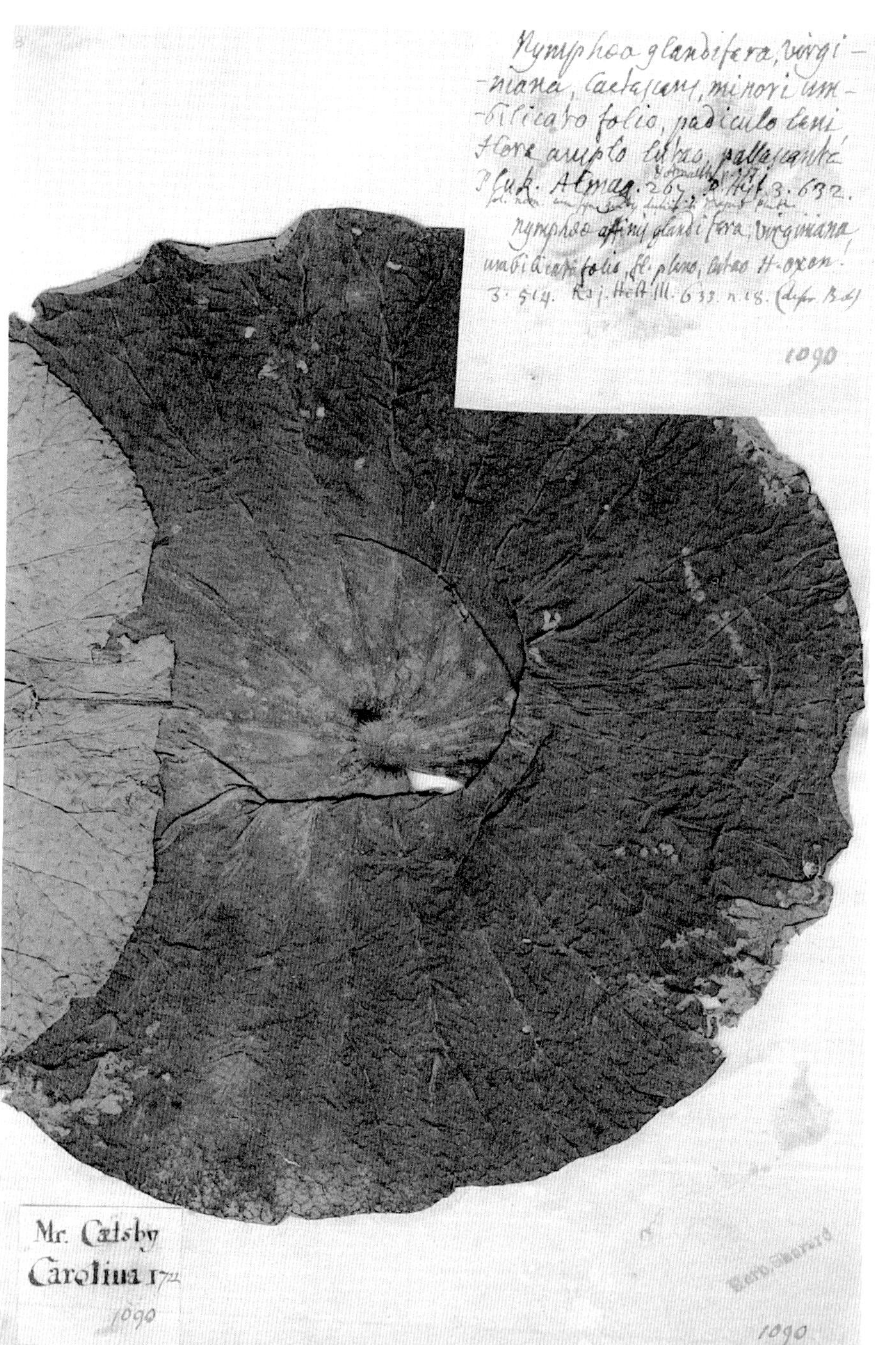

FIG. 2.1. Mark Catesby, *Nelumbo lutea* specimen, 1722. Department of Biology, Oxford University Herbaria.

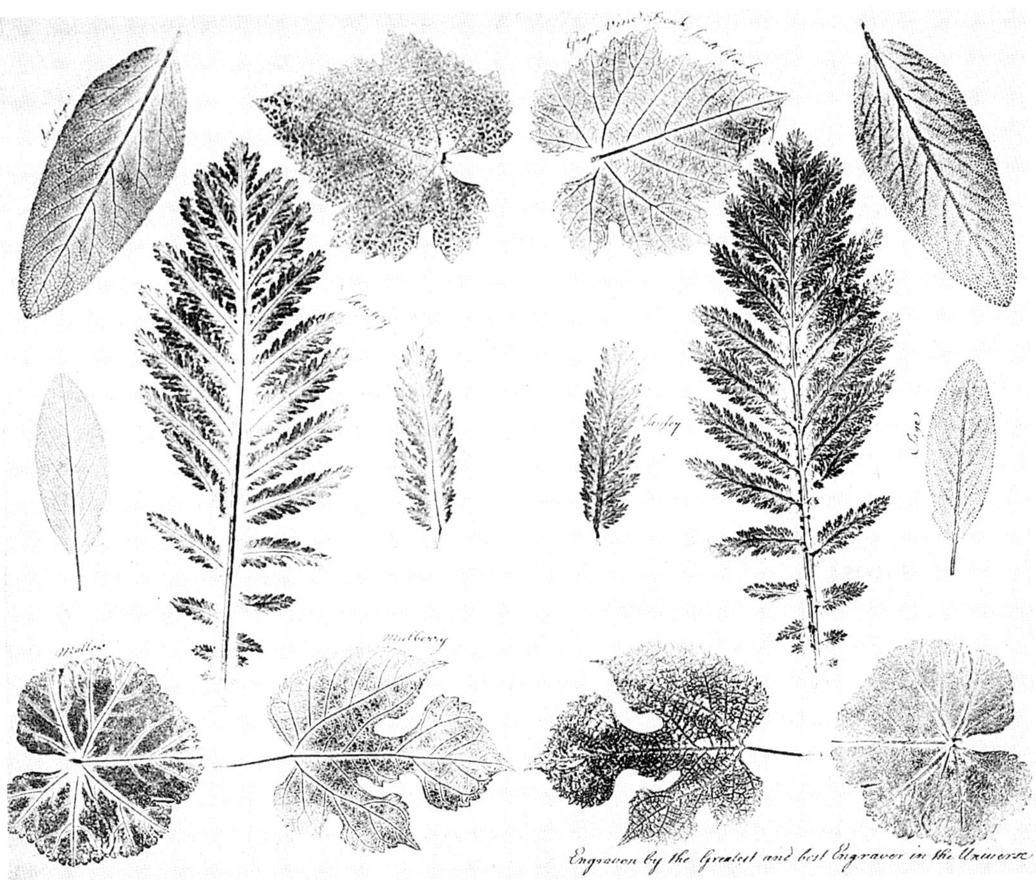

FIG. 2.2. Joseph Breintnall, *Nature Print of Leaves*, 1746. Library Company of Philadelphia.

paper currency as an anti-counterfeit measure.[10] Franklin's transformation of species into banknotes verified their monetary value by cleverly suggesting the "naturalness" of these representations of bullion. Despite their circulation, the notes would always be tethered to a natural origin.[11]

As streamlined and direct as the processes of specimen collecting and nature printing were believed to be, it was understood that much of the natural world failed to yield to these minimally mediated procedures. Moreover, neither specimens nor nature prints were viable options for the printed book. Representation had thus to become more mediated, more distanced from the object represented. Yet for images to *read* as accurate, artist naturalists had to narrow that gap as much as possible through their adherence to particular visual conventions.[12] Seventeenth- and eighteenth-century artist naturalists employed different devices to establish a sense of transparency in their images. Some used an extremely flattened style, echoing the effect of pressed plants or nature prints; others excised the specimen from its original setting, depicting it as though spotlighted, with all its shadows and highlights closely attended to; and still others might situate the portrayed specimen in an agreeable though wholly invented tableau. These different conventions were intended to make the artist's mediating role appear carefully controlled, almost effaced. They presented knowledge as self-evident truth rather than as a process of observation and synthesis.

Flatness was Catesby's preferred technique. Describing his process of "Illuminating" or figuring, he explained that he was not "bred a Painter" and thus had not mastered the skills required to add volume and weight to his forms. He framed this as an asset. "Plants, and other Things done in a Flat, tho' exact manner, may serve the Purpose of Natural History, better in some Measure, than in a more bold and Painter-like Way," he remarked, since this minimized any intervention by the artist that might misrepresent the specimen. It also evoked the appearance of a plant one might find pressed, dried, and enfolded in a sheet of paper.[13] An illustration from the third edition of Catesby's *Natural History*, which purports to depict the *Magnolia acuminata* or cucumber tree (fig. 2.3), reveals how he claimed to follow Woodward's protocol for specimen collecting, even in his images. Here, two of the plant's leaves are portrayed almost entirely flat, drawn after a pressed specimen he received from John Bartram. (He likely did not have a specimen of the flower since the one depicted here is entirely unlike the flower of the *Magnolia acuminata.*) In the appended description, he again followed Woodward, identifying the plant's origin as "the north branch of *Susquehannah* River."[14]

For those naturalists who were trained as artists, other techniques were employed to narrow the distance between the representation and the object in the world. Precise attention to surface, contour, and shadow was encouraged by Royal Society Fellow Robert Hooke in his 1665 opus regarding the visualization of natural knowledge, *Micrographia, or, Some Physiological Descriptions of Minute Bodies Made* (1665). He began this volume with an extended disquisition on the importance of the senses—and particularly of vision—in providing the "Ground-Work" of natural history. He encouraged readers to adhere to "sensible paths," rather than errant wanderings into "invisible notions." Imagination and method were distractions from the task at hand, which required merely "a sincere Hand, and a faithful

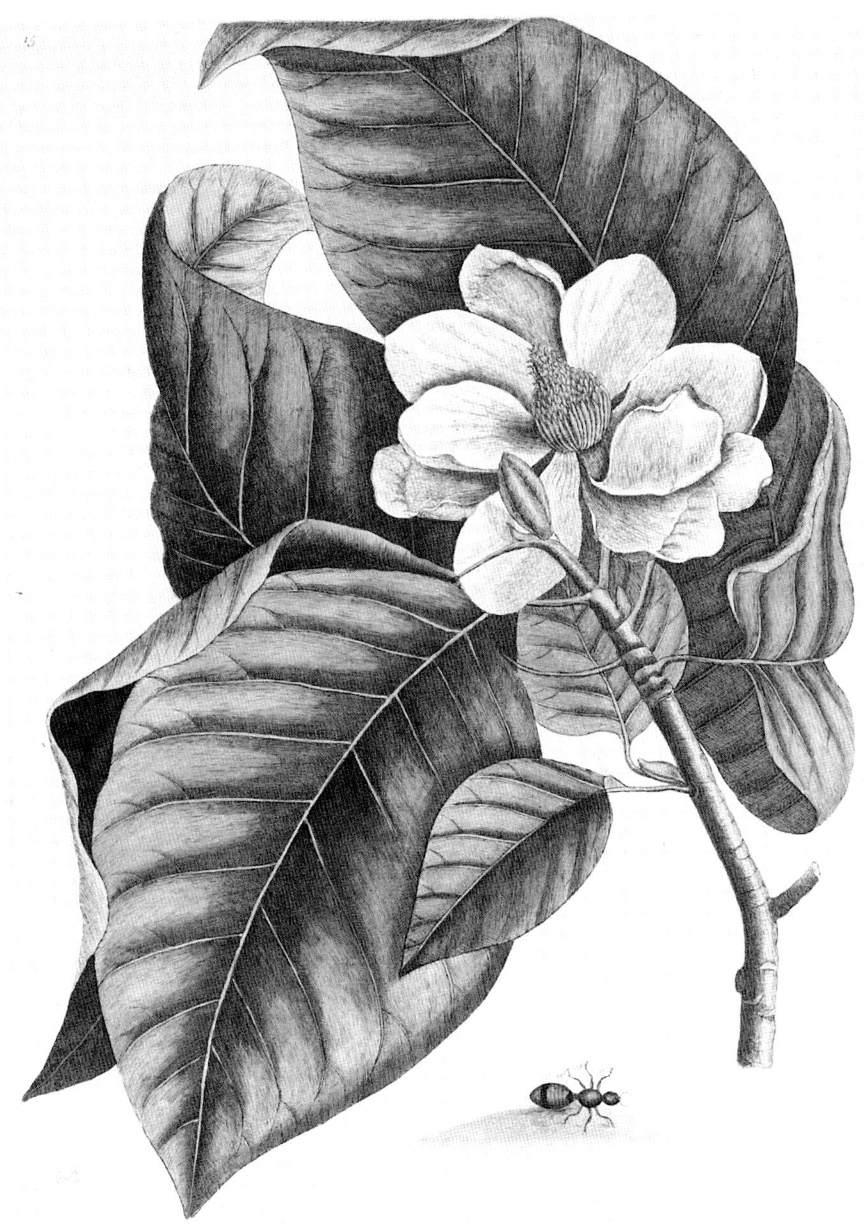

FIG. 2.3. Mark Catesby, *Plate 115: Magnolia Acuminata, or The Cucumber-Tree*. From *Natural History of Carolina, Florida, and the Bahama Islands*, 3rd edition, vol. 2 (London, 1771). University Library, University of North Carolina at Chapel Hill.

Eye, to examine, and to record, the things themselves as they appear."[15] The book's plates evoke this purported artlessness, with many of the specimens portrayed in circular enclosures, as though they were direct recordings of the view through the microscope. In seeming to capture precisely the light and shadow falling across magnified surfaces, the engravings offer an alternative way to close the perceived gap between images and objects. Hooke's *Micrographia* gained new traction in the eighteenth century when the plates were republished as *Micrographia restaurata* in 1745, with revised text by Royal Society Fellow Henry Baker.[16]

Despite the purported eschewal of art or method in the illustrations, both editions have coursing through them a recognition that apprehending the natural world is more complicated than a glance through a lens. Even though the illustrations suggest faithful recordings of the magnified view, the engravings were never direct transcriptions. As Meghan Doherty points out, when one of Hooke's colleagues examined deer hair through a microscope, its substance did not appear to him as spongy, as portrayed in the book's plates, but like a hollow quill. The difference between immediate visual impressions and the illustrations in *Micrographia* is that the latter are composites of different views over time, a record of what Hooke described as the objects' "true form."[17]

Even though the plates were intended to reveal the self-evidence of the natural world, the processes of observation and representation were vexed by the inherent instability of visual appearances. As Hooke observed, *"It is exceeding difficult in some Objects, to distinguish between a* prominency *and a* depression, *between a* shadow *and a* black stain, *or a* reflection *and a* whiteness in the colour." Only through repeated manipulations and observations over time could the observer ascertain the object's "true appearance" and "make a plain representation of it."[18] The process of discovering this true form demanded that the observer examine materials in different positions and under different lighting conditions. Consequently, the visual representation could never be a mere copy of the single, empirical view—instead, it was the product of a time-consuming process of synthesis and rationalization that, according to Janice Neri, transformed fleeting visual impressions into a "quiet, ordered, and comprehensible world."[19]

Despite the extended processes of ordering and integrating sensory data that such illustrations demanded, we can see how the engravings elegantly conceal those processes. The portrayal of a peacock's feather in the 1745 edition of *Micrographia* depicts the barbs and barbules of its vane branching off like "innumerable *Sprigs*" (fig. 2.4), each possessing a "Multitude of bright reflecting

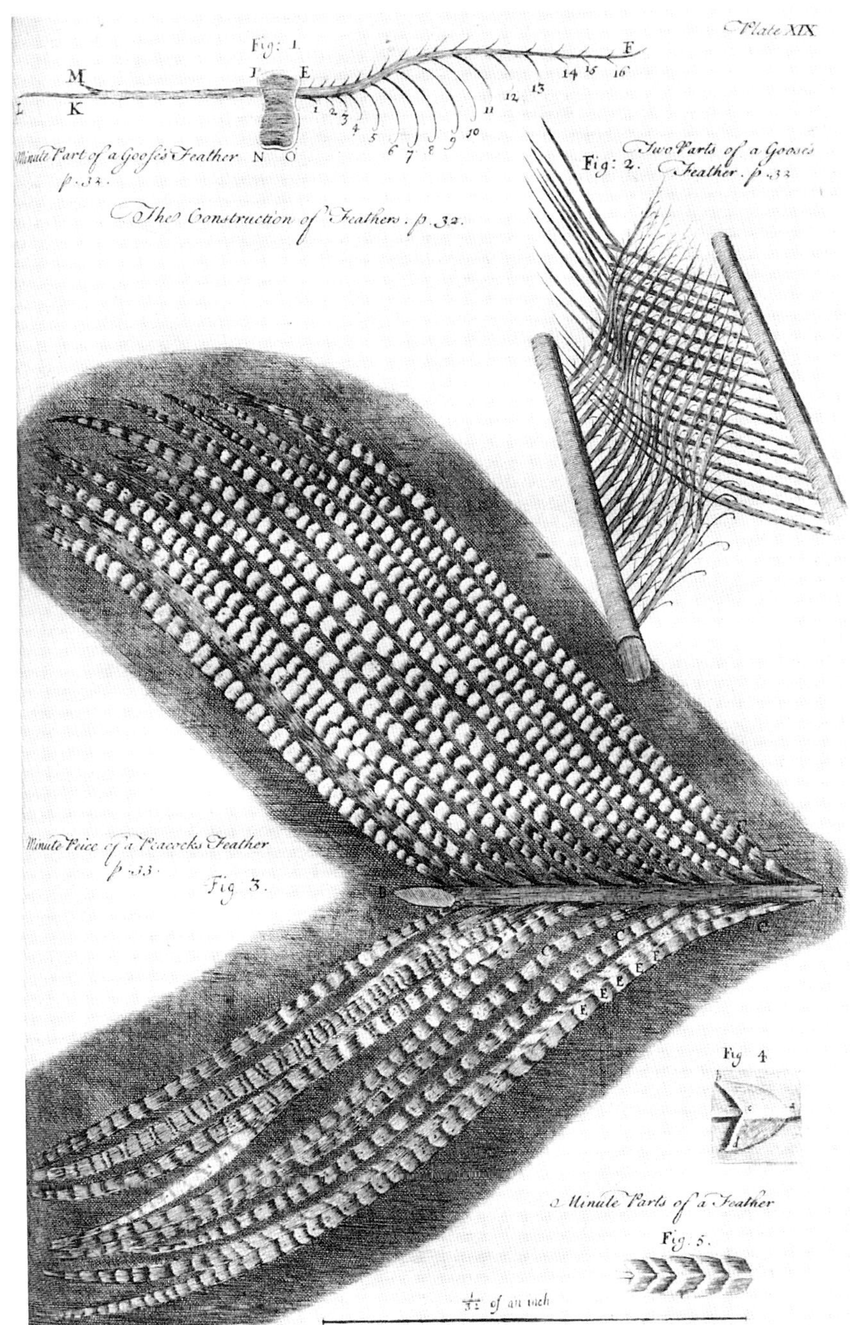

FIG. 2.4. Henry Baker after Robert Hooke, *Plate XIX*. From *Micrographia restaurata* (London, 1745). Bruce Peel Special Collections, University of Alberta Library.

Parts, whose Form and Shape cannot easily be determin'd, since they change continually."[20] Just subtly shifting the intensity of the light completely altered the magnified feather's appearance. Yet the print's carefully patterned rendering belies the disorienting microworld of transparent and reflective surfaces offered by the feather. A bar in the lower right corner denotes scale, and letters mark specific features of the illustration and key them to their verbal descriptions, suggesting a supremely legible natural world.

The composite portrayals we see in the plates of *Micrographia* serve as precursors to the eighteenth century's "reasoned image," a type of representation that demands a particular kind of fidelity to the natural world, what Lorraine Daston and Peter Galison refer to as "truth-to-nature." This fidelity is not simply optical but rational, an instance in which "the eyes of both body and mind [converge] to discover a reality otherwise hidden to each alone."[21] Visual impressions are not simply composited; they are also passed through the ordering grid of a taxonomic system. Where the *Micrographia* plates, amalgams though they are, are still individual portraits, reasoned images are generalized "types" that depict taxonomical concepts rather than individual objects.[22] In the second half of the eighteenth century, these concepts were shaped most profoundly by the Swedish naturalist Carl Linnaeus.

Linnaeus's brief but influential *Systema Naturae* (1735) laid out his sexual system for determining plant classes and orders and, in his first edition of *Genera Plantarum* (1737), he attempted to list all known plant genera arranged according to this system. In the introduction to *Genera Plantarum* Linnaeus included a screed against images—"I do not recommend drawings for determining genera," he wrote, "in fact, I absolutely reject them"[23]—yet images play a significant role in his *Hortus Cliffortianus*, published that same year. In this lushly illustrated folio-sized catalogue of the garden and herbaria of the wealthy Anglo-Dutch financier George Clifford, Linnaeus disposed Clifford's botanical holdings in light of his newly developed taxonomy. He also asked Georg Ehret— the designer of the *Tabella*—to design the book's accompanying plates and to deftly translate his taxonomic method into pictorial form.

The text and image describing the horsebalm or *Collinsonia* (named in honor of Peter Collinson) provides a compelling example of how Ehret made Linnaean plant taxonomy visually manifest (fig. 2.5). The text situates the plant in the class *Diandria*, order *Monogynia*, meaning its flower possesses two stamens and a single pistil. The horsebalm is distinguished from other genera in its class

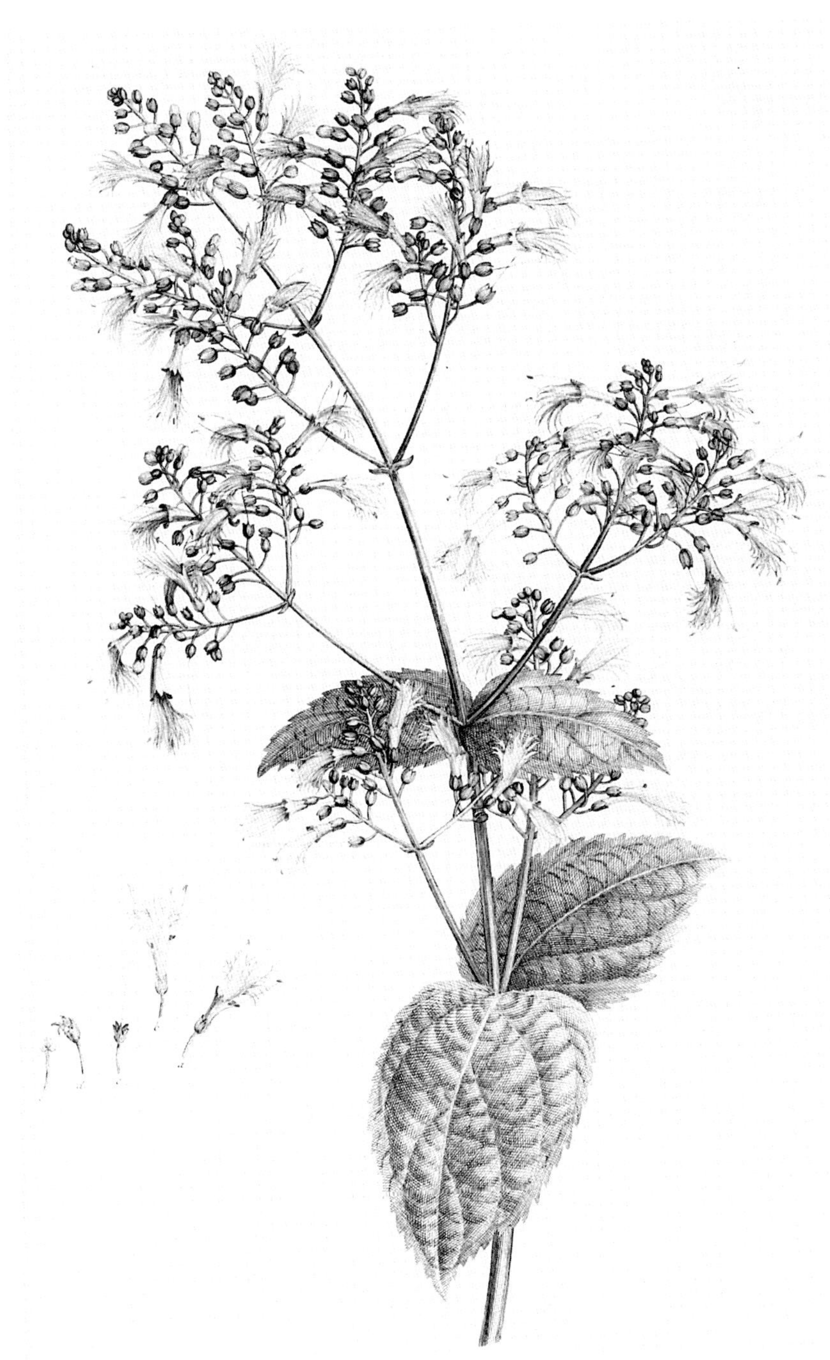

and order by its five-toothed calyx, its small tubular corolla with distinctive fringe, its lack of a seed vessel, and its single round seed. Although the genus was believed to be monotypic, Linnaeus provided its specific characteristics as well. The plant's verbal description in the *Hortus Cliffortianus* includes its fibrous root, its square stem, and most important, its sharply serrated oblong-ovate leaves. Should additional plants of the genus *Collinsonia* emerge, these features would help distinguish it among other species.[24]

Ehret's rendering of the horsebalm maps almost precisely onto Linnaeus's verbal description. He included for reference five isolated figures that show the flower in different positions (*a.* and *b.*), revealing its two stamens and single pistil and highlighting the five-toothed calyx (*c.*) and seed, both fertilized (*d.*) and unfertilized (*e.*). These figures help the viewer locate the characteristics of class, order, and genus in the full branch of the horsebalm, which also includes the specific characteristics of leaf and stem.[25] Although the plant's taxonomic features are carefully delineated to read like a direct visual transcription, this seemingly natural portrayal of the *Collinsonia* is less a portrait than a diagram, in which elements from two tabular plates are transformed into a single veristic image.

The first of these plates is the famous *Tabella* from *Genera Plantarum* (see fig. 1.5); the second is a table of leaf forms (fig. 2.6) from the *Hortus Cliffortianus.*[26] Both plates, in their reductive simplicity, point to an abstract order underlying the surfaces of the world. Although these forms derive from nature, they are distilled down to their essences to convey a truer reality than nature itself.[27] These abstract figures are easily located in Ehret's rendering of the horsebalm— its flowers are embellished versions of figure *B* in the *Tabella*, and its serrated oblong-ovate leaves are adaptations of figure 31 in the table of leaf forms. At the same time, the naturalistic portrayal of these taxonomic characteristics conceals the acts of rational processing, and the engraving deceptively portrays the order of the natural world as supremely self-evident.

Some Linnaean illustrations, as veristic as they may at first appear, took this intellectual abstraction even further. A plant's taxonomic characteristics may not all be visible simultaneously, thus yielding temporally hybrid species. Kärin Nickelsen cites as an example the portrayal of coltsfoot in William Curtis's *Flora Londinensis* (1777–1798), which shows flowers in different stages of development attached to the same rhizome. It also includes the plant's leaves, which only appear after the coltsfoot's flowering stage has ended (fig. 2.7).[28] The image might read as true and accurate, as something the botanist directly

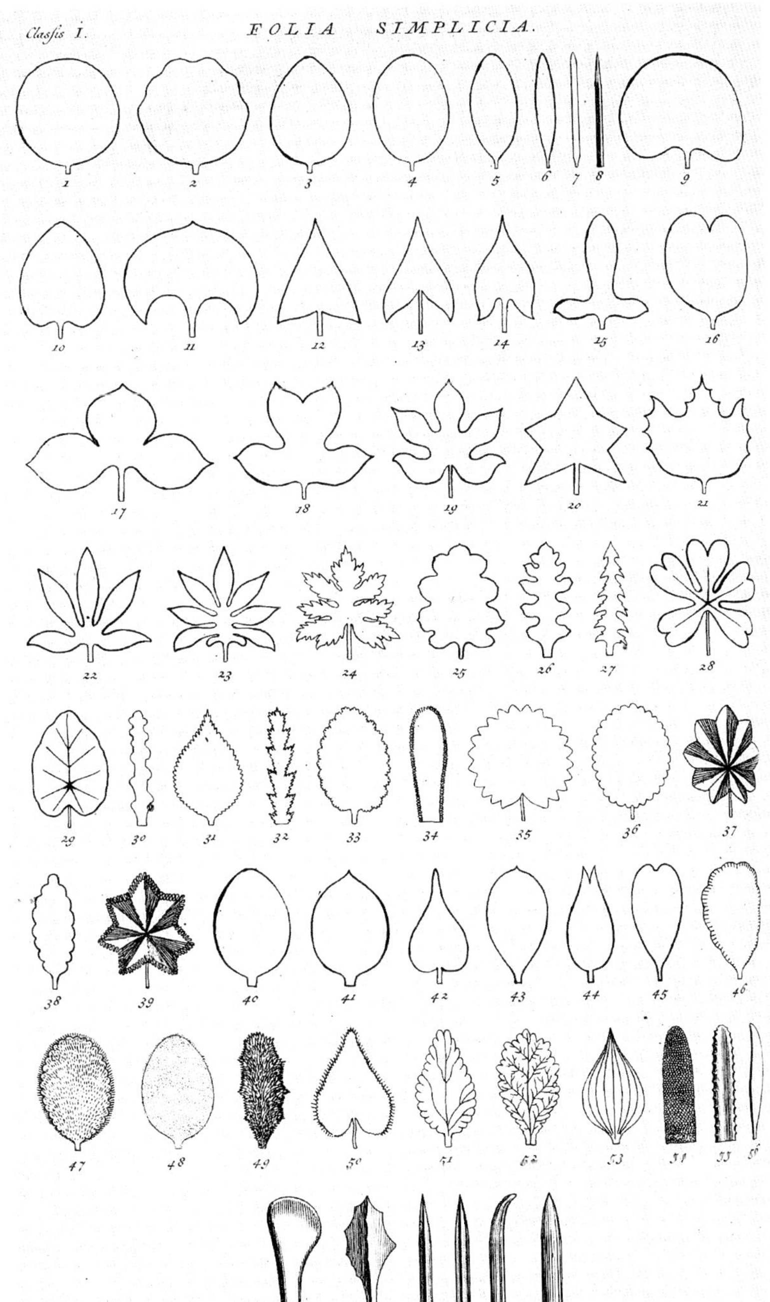

Fig. 2.6. Artist unidentified (Jan Wandelaar, engraver), *Folia Simplicia*. From Carl Linnaeus, *Hortus Cliffortianus* (Amsterdam, 1737). John Hay Library, Brown University (QK98.L77).

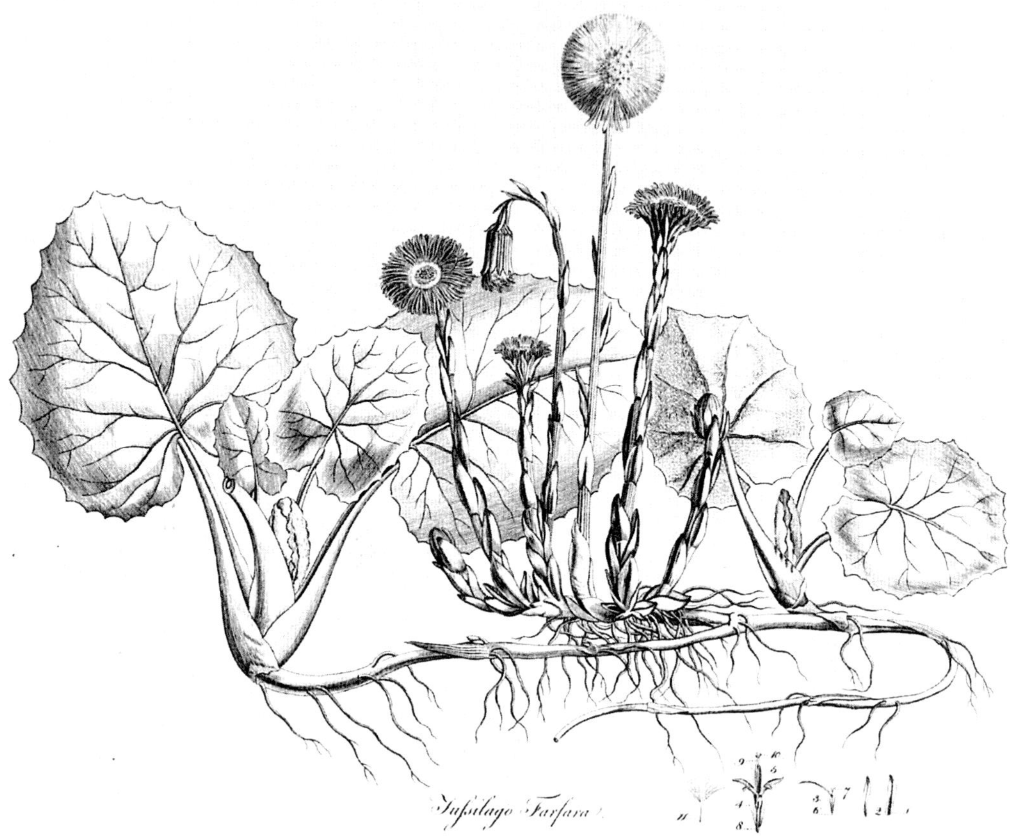

FIG. 2.7. Artist unidentified, *Tab. 60: Tussilago Farfara. Coltsfoot.* From William Curtis, *Flora Londinensis*, vol. 2 (London, 1777–1778). Getty Research Institute, Los Angeles.

observed, yet it is a contrivance, a fictive specimen not to be found in the real world. Although these representational processes were rooted in rigorous empirical observation, such observations were then passed through a conceptual grid. Observations were assessed, winnowed, and combined into solitary figures that were, ultimately, inventions.

If Catesby flattened the specimens he portrayed, and Hooke and Ehret isolated their objects and captured their volume and contour in careful chiaroscuro, other naturalists such as George Edwards attempted to construct cozy vignettes to enhance their illustrations' sense of accuracy. Just like the images of Catesby, Hooke, and Ehret, Edwards's are likewise intellectual constructions, yet he emphasized pose and environment in an effort to make his images appear lifelike. In the preface to his *Natural History of Uncommon Birds* (1743–1751), Edwards wrote, "I have made the Drawings of these Birds directly from Nature, and have, for variety's sake, given them as many different Turns and Attitudes as I could invent," explaining that he had also elaborated the "Grounds" with additional flora and fauna. Although the intent was to avoid the unpleasant sameness of most ornithological illustrations, he was concerned with more than just visual interest. His inventions were also meant to make his etchings "natural and agreeable," thereby transforming the book's plates into scenes that viewers might encounter.[29]

Of course, Edwards's text makes clear that the depicted scenes are not always *possible*—they often bring together fauna culled from different geographical regions—yet they still appear *plausible*. It seems not unlikely that two black-bellied hummingbirds, possibly mates, might alight on a tree stump while a fat fly buzzes past (fig. 2.8). The text, however, indicates that these creatures would never cross paths in real life, since the birds are from the American tropics and the fly from Amboyna. Just as the birds and fly would not encounter each other in life, neither would they in death: the male specimen was owned by James Theobald, the female by Taylor White, and the Ambonese fly was held in the collection of Joseph Dandridge, founder of one of London's early entomological societies.[30]

The invented environments Edwards crafted for his birds may differ from the shocking flatness of Catesby's illustrations or the isolation and decontextualizing of specimens found in the work of Hooke and Ehret, but the illustrations share similar intent. In all of them the artist endeavored to present the natural world as a set of visually self-evident facts, obscuring the manual and mental effort required for the discovery and synthesis of those facts. In their illustrations, these artist naturalists transformed a confusing array of visual impressions into smooth seamless wholes. The sense of accuracy in these images turned on the fiction that they were transparent windows onto the world, offering direct unmediated views. The flattened bloom and leaves of the cucumber tree evoke a pressed specimen; a microscope presents a tidy miniature world; the horsebalm reveals its class, order, and genus; the perched hummingbirds and buzzing fly suggest a scene clipped from nature. In these images, accuracy depended on transparency, but transparency required a concerted act of concealment. As we shall see, Bartram's drawing of the American lotus—with its inconsistent scale and multiple points of view—emphatically rejects this dissimulation.

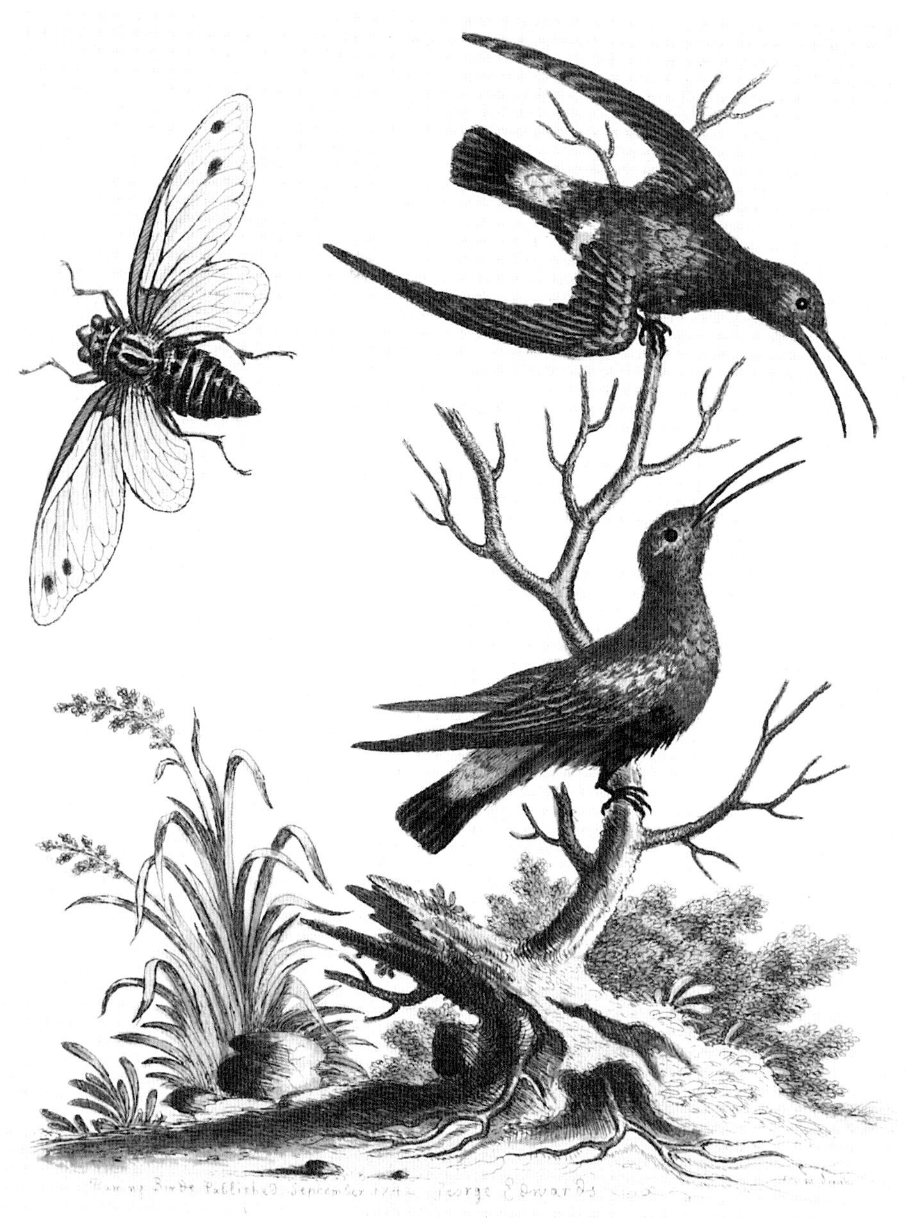

FIG. 2.8. George Edwards, *Plate 36: The Black-belly'd Green Huming Bird*. From *A Natural History of Uncommon Birds*, vol. 1 (London, 1743). Digital Library for the Decorative Arts and Material Culture, University of Wisconsin–Madison.

The True Colocasia

In his letters to the Bartrams, Collinson referred to the primary subject of William's 1767 drawing using some variation of *faba aegyptica* or *colocasia*, the Latin and Greek names for the American lotus's Old World cousin.[31] Seventeenth- and eighteenth-century merchants and mariners from Europe had observed the sacred lotus (*Nelumbo nucifera*) in China, Java, and the southwestern coast of India, but its purported origin was in northern Africa, as attested in Dioscorides's first-century natural history, *De materia medica*. Believing the American lotus (*Nelumbo lutea*) to be a variant of this species, Collinson and his fellow plant enthusiasts were understandably eager to get a better sense of this botanical ligature among the foreign lands of Asia, Africa, and America.[32] Admittedly, the body of knowledge surrounding the lotus was confused and confusing; classical texts, devoid of illustrations, contradicted one another and were confounded by modern interpretations. If the plant's exoticism made it an exciting subject for Collinson, then the literature on the plant, stuffed as it was with inaccuracies and misattributions, made it an especially compelling topic for William Bartram as he challenged the standards of natural history representation.

William would have been familiar with the lotus's tangled history. Generations of European naturalists had struggled to ascertain its genus, and it had for many years been considered a type of *Arum*. Yet even as early as the mid-seventeenth century that designation was placed in doubt: in his *Theatrum botanicum* (1640), John Parkinson grouped the plant among the *Arums* despite—as he readily noted—being entirely unlike any species of that genus.[33] Its odd placement in Parkinson's text was intended to refute the common misidentification of the lotus or the "true *Colocasia*" with the Egyptian arum or culcas (now commonly referred to as taro). Although Parkinson hazarded no suggestion of how the plant should be classed, he knew it was not an *Arum*, and he juxtaposed it with species of that genus to underscore its difference. "Because the *Egyptian Arum*, hath beene so much mistaken by many writers that have called it the true *Colocasia* of *Dioscorides* and *Theophrastus*," Parkinson assured the reader, "let me here shew you in this place, the description of the true *Colocasia*."[34]

Parkinson's text is a determined effort at clarification. He first summarized classical descriptions of the colocasia (I use "colocasia" here to refer to the lotus since this was the term adopted by early modern authors), which possesses large rounded leaves, double-flowered blooms, and a ciborium-shaped fruit. The fruit's top should be pocked with cells, each cell holding a round seed or nut. The Egyptian arum, he pointed out, does not at all align with this description: it has long arrow-shaped leaves, a thick stalk, and flowers composed of "huskes or hoses . . . [with a] pestell or clapper in the middle" (that is, spathes or leaflike hoods with a central spadix, or spike inflorescence) and a great bulbous root with fibrous protuberances.[35] The only similarity between the lotus and the Egyptian arum, he noted, was that the root of both plants served as food. In explaining the confusion of the two species, Parkinson attributed it to a semiotic problem among sign, signified, and referent that was engendered by the plants' physical and visual inaccessibility to European authors. With neither the living plants nor images to guide them, early modern interpreters struggled to fit known species to the descriptions offered in classical botanical texts. Despite bearing a very different fruit and flower from those that classical authors attributed to the colocasia, the Egyptian arum was, Parkinson conceded, the plant that seemed "most nearly [to] approach thereunto." The daily use of taro for food "among those Nations of *Egypt*, *Syria*, *Arabia*, and *Affricke*," and its regional name of culcas—so close to colocasia—made it a promising candidate.[36]

Parkinson reserved special distaste for the sixteenth-century Italian physician and naturalist Pietro Andrea Mattioli, who had further confused the matter in his commentaries on Dioscorides (fig. 2.9). Mattioli had made much the same argument as Parkinson, that the true colocasia of classical texts was not to be confused with the Egyptian arum, and his assessment was supposedly based on direct visual evidence. Mattioli claimed to have seen an actual specimen of the colocasia brought to Trent along with other exotics from Egypt and Syria and had been given a drawing of the Egyptian arum by an ambassador to Constantinople.[37] And yet, despite his visual observations, the illustration he provided of the colocasia accords with no known species.

Parkinson opined that Mattioli's figure was "moulded from his owne imagination, and not from the sight of any plant growing *rerum natura*, to make it answer the description" given in classical texts.[38] In fact, Mattioli's illustration fails even in that regard, portraying few of the features described by Dioscorides or Theophrastus. He seems to have taken the general form of an arum and added some of the elements Dioscorides attributed to the colocasia. Arums typically have pointed leaves and a central inflorescence that bears the plant's flowers and, later, its fruit. Mattioli's image depicts a spadix after the flowers

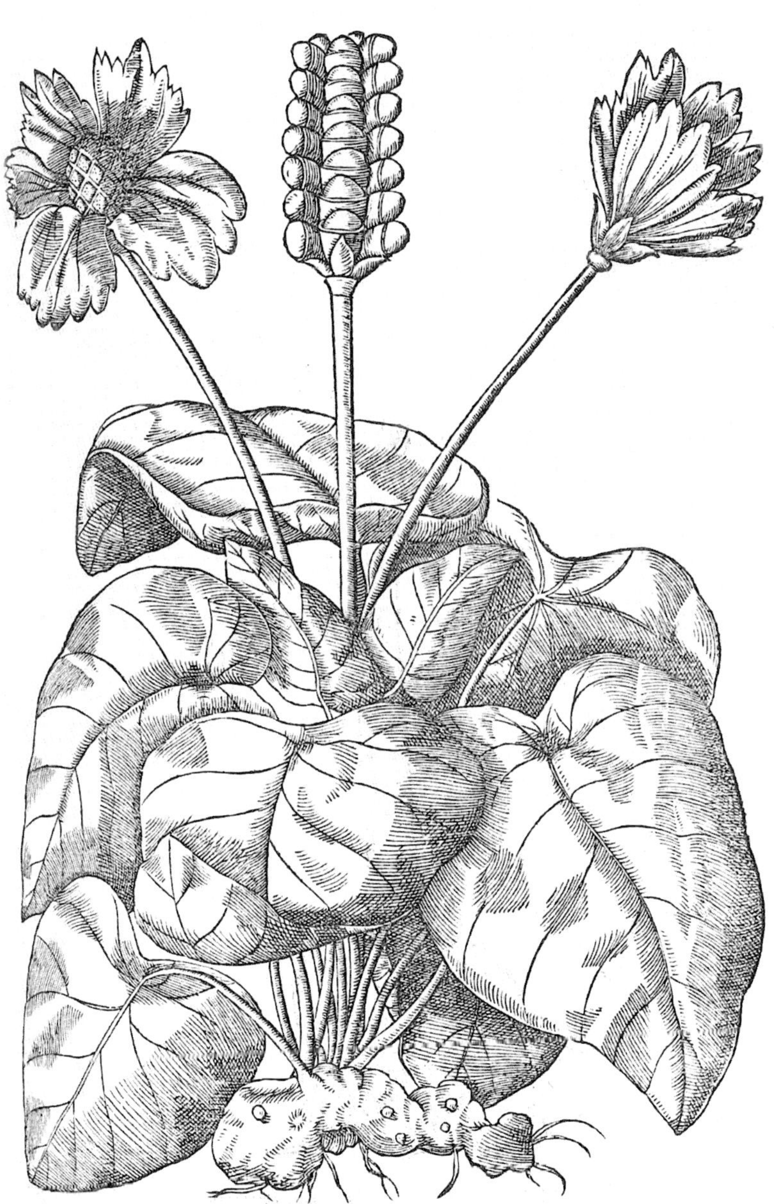

Fig. 2.9. Giorgio Liberale, *Fava d'Egitto*. From *I discorsi di M. Pietro Andrea Matthioli . . . de Pedacio Dioscoride Anazarbeo, della materia Medicinale* (Venice, 1568). Getty Research Institute, Los Angeles.

have fallen away and the berries have ripened. Yet he also included a second flower-and-fruit combination on the plant, featuring on either side of the spadix a poppylike bloom, one of which bears a cone fruit in its center. Mattioli's figure is a curious botanical fusion that seems to belie his claim to firsthand observation.

Parkinson himself made no claim to have seen the lotus or colocasia in life; instead, he scoured the existing botanical literature for textual and visual references that aligned most closely with classical texts. Carolus Clusius's description of the plant and illustration of its fruit in his *Exoticorum libri decem* (1605)—a compendium of unfamiliar life forms—proved, in Parkinson's view, most reliable. Clusius had seen a specimen brought to Amsterdam from Java by the merchant Emanuel Zweert, and his description accorded not only with those of the classical authors but also with observations offered by his contemporaries, including the missionary Justius Heurnius and the English traveler William Fincham.[39] Parkinson, likely because of his lack of firsthand experience with the species, offered no illustration of the full plant but chose instead to reproduce Clusius's figure of the fruit.

Despite Parkinson's efforts to disentangle the confusion regarding the colocasia, he was not wholly successful, and misunderstandings about the plant extended well into the eighteenth century. The ongoing use of "colocasia" to refer to the Egyptian arum and the continued inaccessibility of the lotus likely enhanced the term's referential instability. The sacred lotus was not cultivated in Britain until Joseph Banks introduced a viable specimen in the 1780s, and the American lotus was not observed in British gardens until 1824.[40] Even Philip Miller of the Chelsea Physic Garden perpetuated the error, cross-referencing the lotus with the *Arum maximum AEgyptiacum* in his *Gardeners Dictionary*, which in turn influenced Collinson.[41] In his first letter to the Bartrams regarding the lotus, Collinson wrote somewhat uncertainly, "I always thought the Colocasia was a Species of Arum It is so esteemed by the moderns," suggesting that interest in the plant had done little to correct misconceptions.[42]

William Bartram, who was able to observe the American lotus growing in its habitat, seems not to have been similarly misled. Throughout his writings, he used "colocasia" to refer only to the lotus, modifying it to "Arum colocasia" to describe the Egyptian arum. The plant's uncertain origin and knotty history seems to have transformed it into an emblem of representational error for Bartram, making it an apt subject for his interrogation of a supposedly self-evident natural world. Like Hooke, William recognized the inherent instability of visual impressions. As a result, his observational process did not merely register the visual *appearance* of nature but, rather, investigated its materiality through a process of somatic engagement. The optic is brought level with the haptic in his natural history, and sight and touch become equally important: his movement in and through the natural world depended on vision, yet it was only through the combination of his senses that he could truly assess that world.

In Bartram's published account of his botanizing trip through the American South, *Travels through North & South Carolina, Georgia, East & West Florida,* it is evident how important the materiality of nature was to his understanding of it. A hallmark of late eighteenth-century nature writing, the book exerted significant influence on such writers as William Wordsworth and Samuel Coleridge.[43] Bartram's appeal to the Romantics is not difficult to understand since *Travels* is rife with aestheticized descriptions of enchanting vistas typical of late eighteenth-century travelogues. Yet Bartram was not solely in search of a view; instead, he was focused on parsing the illusory from the real. While traveling along the St. Johns River in the 1770s, he described an "enchanting" but implausible scene in which the waterway seems but "a grand avenue . . . [that] elude[s] the exact rules of perspective and appears an equal width." This visual experience at once inspired and was ultimately corrected by a kinesthetic one, as he traveled further along the St. Johns and into the great Lake George. Passing through the landscape, he discovered that the river gradually widened over the course of its two-mile stretch, such that it appeared not to recede in perspective.[44] Here vision supplied the origin of his exploration yet did not conclude it; he instead relied on his bodily engagement with space.

Travels offers valuable insight into Bartram's understanding of empirical observation and representational conventions and how that understanding played out in his drawing for Collinson. In the book's second part, which centers on his excursion through the Floridas, Bartram described a distant cluster of lotuses that evoked "a delusive green wavy plain." The prospect altered as Bartram approached, and he began to augment his visual perception with additional sensory impressions. The unified plain broke into a series of "gay flowers, waving to and fro on flexible stems, three or four feet high," which he described as "double as a rose, and when expanded . . . seven or eight inches in diameter, of a lively lemon yellow colour." He then turned his attention to the fruit, "a large truncated, dry, porous capsule, its plane or disk regularly perforated," which contained the plant's seeds, "each . . . an oval osseous gland or nut, of the size of a filbert." His visual consideration of the plants gave way to his physical examination of the seeds, which he described as hard and inedible when fully ripe, but "sweet and pleasant eating" when young. Here Bartram's shifting optical impressions led to haptic and even gustatory ones, as he engaged in a full bodily assessment of the plant.[45]

Bartram's deployment of his other senses to correct his visual perception tracks back to philosopher John Locke's theory of sensory heterogeneity. A rejection of the Cartesian model of the *sensorium commune*, this theory maintained that information derived from one sense could not be transferred to another. One would, for example, be incapable of recognizing something by sight that was previously known only by touch. As a consequence, observers must bring all their senses to bear on an object in order to understand it. This model of the senses was widely disseminated during the eighteenth century through such popular texts as Noël-Antoine Pluche's *Spectacle de la Nature* (first French edition, 1732). Pluche indicated that the correct way to ascertain an object—what he called the "true Philosophy"—was to collect as many different sense impressions of it as possible.[46] To illustrate his point, Pluche invented a scenario that drew on the active transatlantic trade in botanicals. In an impish twist, he made René Descartes his protagonist: "Present to *Descartes* a Pine Apple freshly cut off from the Stalk, perfectly ripe. Desire him to examine the inward Frame of that Fruit, which is but just begun to be cultivated in *Europe*, and from that bare Inspection to tell you what the Taste of it must be. . . . *Descartes* will never find out the Savour in his Reason, nor even in the Concurrence of the Elements and Vessels of the Fruit itself, after having analyzed and dissected it; his Palate alone can and will let him into that Secret."[47]

As an artist naturalist corporeally embedded in the world that he studied, Bartram interrogated the privileging of vision in scientific investigation, as well as the effacement of the colonial artist naturalist. In particular, he called into question the idealized transparent transference of the object in the world to the image on page. In an oft-quoted passage from his *Travels* he illustrated how profoundly misleading this sort of transparency could be. Describing a deceptive view into the crystalline waters of an East Florida spring, Bartram observed, "This amazing and delightful scene, though real, appears at first but as a piece of excellent painting; there seems no medium, you imagine the picture to be within a few inches of your eyes, and that you may without the least difficulty touch any one of the fish, or put your finger upon the crocodile's eye, when it really is twenty or thirty feet under water."[48] As with his prospect of the St. Johns or his distant view of a field of lotuses, Bartram delighted in the scene, but his attempt to check his visual perception via touch altered his understanding. Although the fauna he perceived appeared to hover within reach, his haptic sense discovered this fiction, and he recognized that the reptile that seemed so very close at hand was deep below the water's surface.

According to Michael Gaudio, this passage indicates the tension between visibility and invisibility, self-evidence and obscurity, in Bartram's natural history. Gaudio cites a passage from Bartram's meditation on human nature that reveals the naturalist's deep distaste for dissimulation: "of all Our passions, sentiments or affections Dissimulation, is the most Mischievous, & indignant & I believe that no person more highly approves of simplicity & sincerity than myself."[49] Interestingly, Bartram appears to have observed this moment of supposed crystalline transparency at an East Florida spring as a form of visual dissimulation, concealing as much as it revealed.[50]

Considered in this way, Bartram's framing of the illusory scene as "an excellent piece of painting" reads more as indictment than praise.[51] Although it would be difficult to determine whose painting provoked the metaphor, John Singleton Copley's canvases serve as a compelling model. Centered in Boston during the 1760s and early 1770s, Copley was known throughout British North America as the colonies' finest native-born painter. Copley's precise visual descriptions of flesh and fabric have been understood as participating in the same empiricist project that motivated eighteenth-century naturalists and explorers. By the time Bartram wrote his *Travels*, he likely knew Copley's portrait *Mr. and Mrs. Thomas Mifflin* (1773; fig. 2.10), since Mifflin—a prominent Philadelphia merchant and politician, as well as a fellow alumnus of the Academy of Philadelphia—was the book's dedicatee.

The portrait is representative of Copley's American style, which delineates every detail in a master study of material surface, capturing the textural differences of the sitters' clothing, the crushed velvet of the upholstery, the slick surface of the table, and the skeins of thread stretched on Mrs. Mifflin's loom. And yet, despite Copley's unwavering attention to the materiality of the objects he portrayed, the painting is as much a project of concealment as revelation. Viewers are intended to see not linseed oil mixed with pigment and applied to canvas but, rather, to see *through* it, to the fabric of the sitters' habiliments and the mahogany of the table. It is this seeming lack of medium that is so at

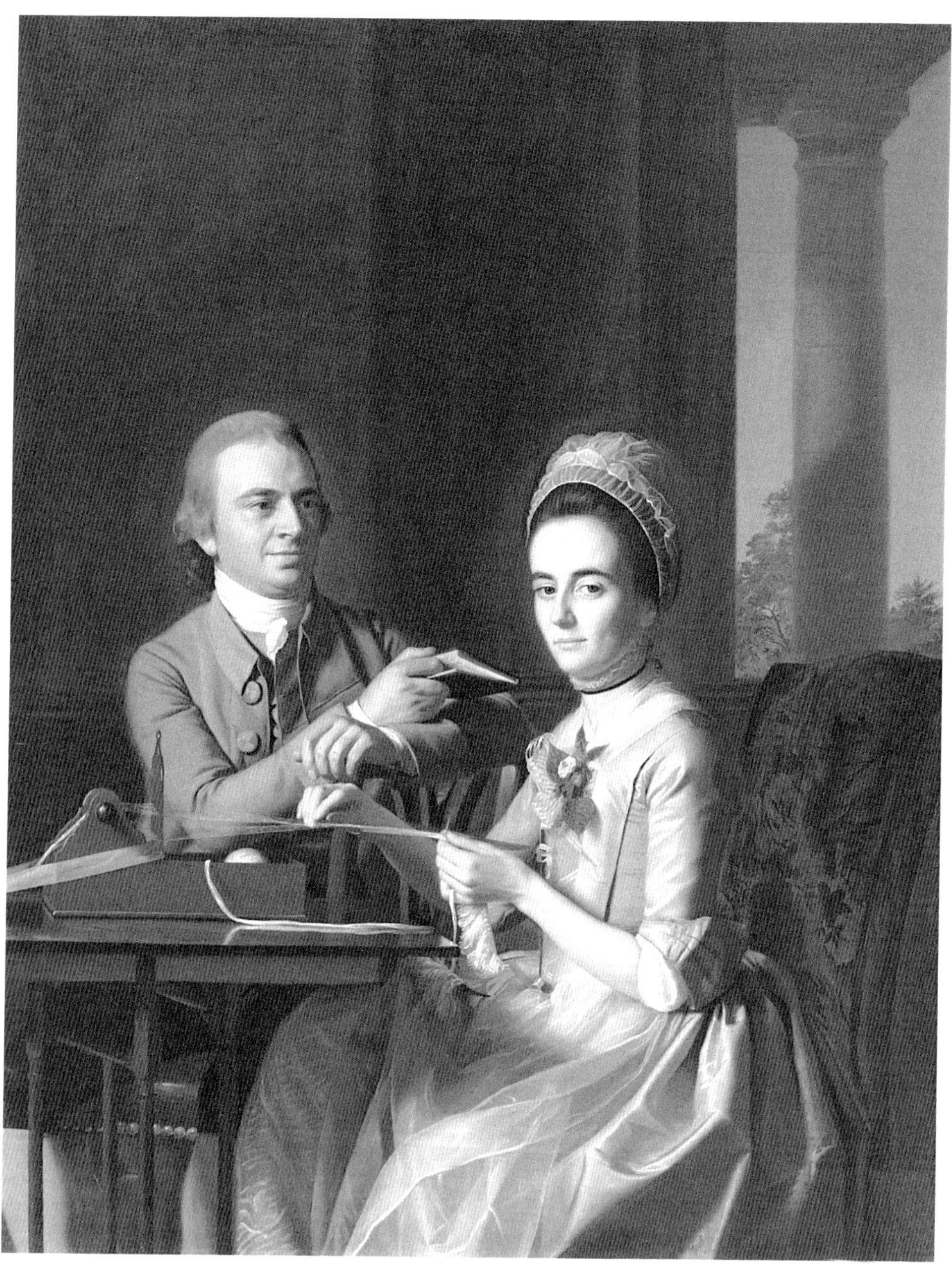

FIG. 2.10. John Singleton Copley, *Mr. and Mrs. Thomas Mifflin (Sarah Morris)*, 1773, oil on ticking. Philadelphia Museum of Art.

issue in Bartram's reflection on picturing, where transparency conceals distances and deceptively presents one surface as another. The acclamatory adjectives in Bartram's passage bobble on an undercurrent of disappointment. Visual representations that attempt to create optical analogs of the world provide only false testimony.[52]

The failure of such images to convey truth demanded that Bartram pursue a separate tack. It would seem that any form of visual representation would fall short of articulating his investigations of nature, but as his descriptions of his southern travels make clear, enchanting prospects were central to spurring his pursuits. He observed the natural world before him, composed as a picture, while refusing to take its visual appearance alone as evidence of itself. Instead, he trekked into and through the view in order to test its appearance against its material presence. Rather than stand as a simulacrum of the natural world, Bartram's drawing of the lotus instead endeavors to capture his very process of discovery by providing an initial prospect, followed by a sense of exploratory movement.

A Curious Performance

Bartram delineated his process of discovery in his lotus drawing by creating an unexpected conjunction of different representational conventions. The drawing's oddly flat lotus leaf evokes the simplicity of the botanical specimen or nature print, while the flowers mimic the Linnaean format established by Ehret. These elements are all presented in their habitat and bracketed on the left by a great blue heron, a dragonfly, and a Venus flytrap. The constructed environment seems akin to Edwards's pleasing tableaux, except the additional flora and fauna do not contribute to a plausible setting for the plant—leaf, flower, fly, and bird jostle together in a disjointed pictorial space. One might read this fragmentation as indicative of Bartram's inability to synthesize, but it also serves to counter the conception of a self-evident natural world. By cobbling together different conventions for conveying nature, Bartram's drawing does not present knowledge as a seamless self-evident whole but, rather, encourages the viewer to trace his construction of it. Through this unintegrated accretion, the drawing uncovers the falsehoods of natural history representation and its operative principle of effort-effacing transparency.

Perhaps most arresting in Bartram's composition is the flat leaf floating on the surface of the water, surrounded by three blossoms and two towering leaves.

While the flowers' staggered and overlapping forms imply a three-dimensional space, the floating lotus leaf does not participate in this spatial construction. It tips forward, as though pressed against the surface of the paper, mimicking a botanical specimen or an inked and printed leaf. At the same time, Bartram's drawing of the American lotus gives the flower its full due, in accord with Linnaean dictates, and presents a bud, an open blossom, and a fully blown bloom. The bud offers a clear view of the calyx, while the open blossom highlights the corolla's double row of petals. The fully blown bloom—depicted at the instant before the petals fall away—reveals the flower's many stamens (as corresponds to its Linnaean class, *Polyandria*) and the lotus's distinctive fruit, pocked with seeds.

Even as Bartram used the specimenlike leaf and Linnaean flowers to convey his material and conceptual familiarity with the lotus, the way he incorporated them into a single tableau ultimately undermines any sense of transparency. Bartram situated them in a pond, with a great blue heron perched at its edge. The bird's posture suggests a moment of narrative suspension, as it tracks the movement of a fish through the water. This vignette not surprisingly calls to mind George Edwards's invented environments, since it is, in fact, a quotation from volume 3 of *Uncommon Birds* (fig. 2.11).[53] The wings held gently away from the heron's body, its lifted left foot, and the position of the swimming fish are all deftly copied from Edwards. Yet Bartram's citation does not create a more natural and agreeable environment for the lotus; instead it reinforces the drawing's sense of disjunction. Where the flat leaf disrupts a tentative construction of depth, the great blue heron undermines it entirely. Bartram could have chosen any number of species from Edwards's publications to augment his portrayal of the lotus, yet he selected a strikingly large bird, while shrinking it to a small addendum at the edge of his composition. It looks in danger of being enveloped by the leaf behind it, even though the leaf's natural diameter of twelve to sixteen inches is only one-third of the heron's projected height. Because of the drawing's disorienting inconsistencies of scale and alternations between surface and depth, the portrayed objects do not read as part of a legible perspectival scheme. The composition offers a spatially indeterminate world that quavers between flatness and fullness.

The unsettled sense of space in Bartram's rendering of the lotus evokes William Hogarth's frontispiece for *Dr. Brook Taylor's Method of Perspective Made Easy* (1754), a book held by the Library Company of Philadelphia. The text was written by Joshua Kirby, a colleague of Hogarth's, as an intended aid to artists.

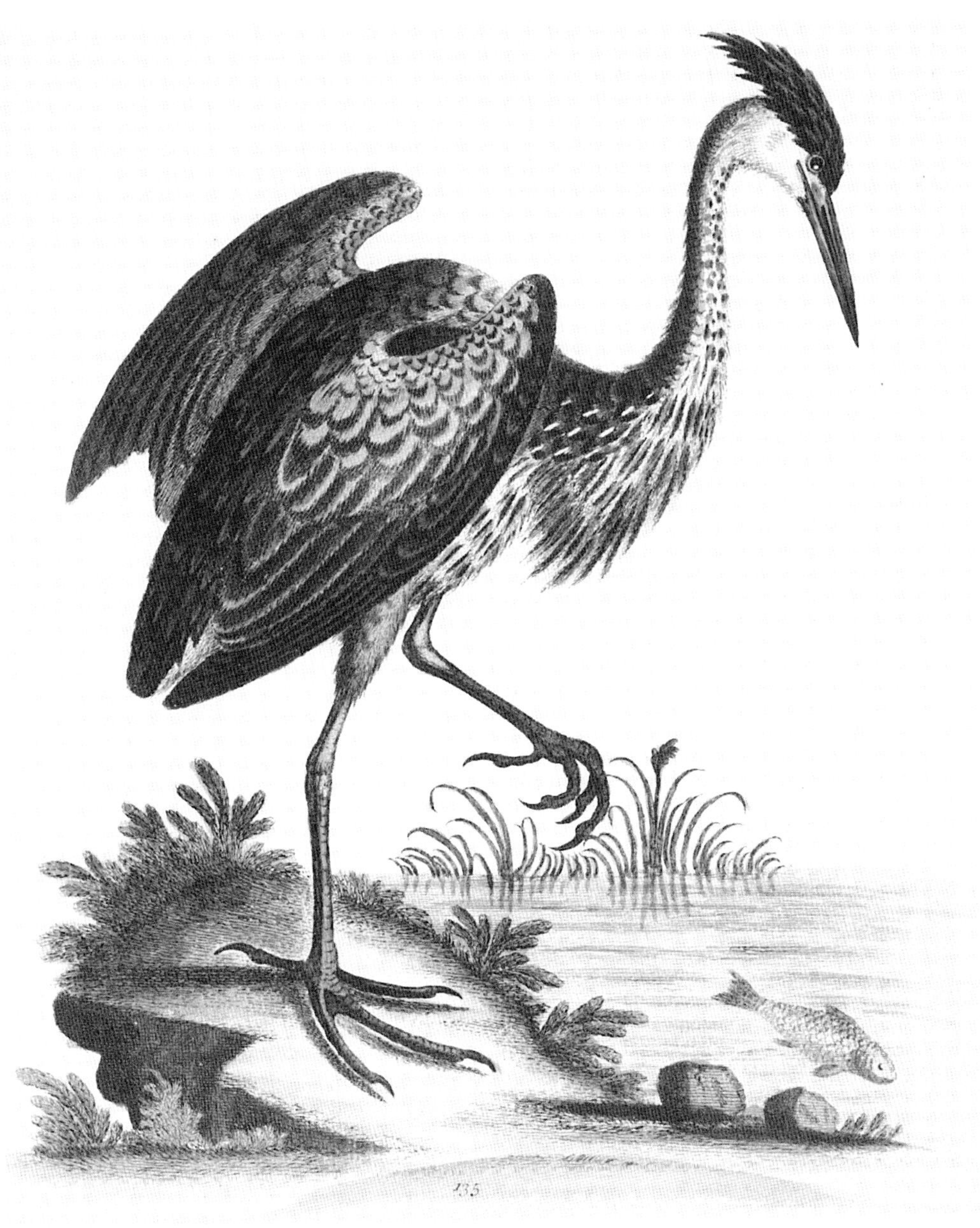

FIG. 2.11. George Edwards, *Plate 135: The Ash-colour'd Heron from North-America*. From *A Natural History of Uncommon Birds*, vol. 3 (London, 1750). Digital Library for the Decorative Arts and Material Culture, University of Wisconsin–Madison.

In the book's preface, Kirby averred that "to draw a good Picture is to draw the Representation of Nature, as it appears to the Eye; and to draw the Perspective Representation of any Object, is to draw the Representation of the Object as it appears to the Eye."[54] Hogarth's frontispiece operates in a satirical vein, portraying a cockeyed world of spatial incoherence, as walls abut each other at awkward angles, distant objects overlap ones in the foreground, and forms present more than one side simultaneously (fig. 2.12). "Whosoever makes a DESIGN without the Knowledge of PERSPECTIVE will be liable to such Absurdities," the caption for the frontispiece warns. Yet perspectival rendering, as articulated by the diagrams and guides in Kirby's text, functions similarly to Linnaean taxonomy by providing an overarching rational structure that suppresses certain details while highlighting others.

Bartram seems to have rejected such a rigorous diagramming of space in the construction of his lotus drawing and focused more specifically on Collinson's request for a "picturesque figure."[55] In the 1760s the picturesque most commonly referred to a Claudean landscape composed of interlocking wedges of color and tone that yielded an agreeable gestalt and a gentle movement in and through the composition. Hogarth shared this view but maintained that the picturesque need not actually cohere into a legible representation of three-dimensional space. In his 1753 aesthetic treatise, *The Analysis of Beauty*, he observed that an image representing no particular scene but composed of "lights and shades only, properly disposed . . . might still have the pleasing effect of a picture."[56] Similarly, the theorist Daniel Webb suggested that the picturesque consisted largely of the proper disposition of forms that yielded some larger, if abstract, pattern. Although Webb would turn these compositional ideals explicitly toward a convincing portrayal of volume and depth, the picturesque was fundamentally understood to be the visually pleasing arrangement of lights and darks across a surface.[57]

"When lights and shades in a composition are scattered about in little spots, the eye is constantly disturbed," Hogarth observed. He directed artists to carve the "composition into three or five parts, or parcels" to establish "that variety which entertains" rather than disturbs the eye.[58] Bartram appears to have followed these directives closely in his drawing for Collinson. In the striated bottom band, encompassing the lower third of the composition, Bartram provided a dark, contrasting background for the lotus leaf and the delicately balanced great blue heron. In its top register he inverted this light-against-dark arrangement by leaving the background untouched and figured against it the three flowers

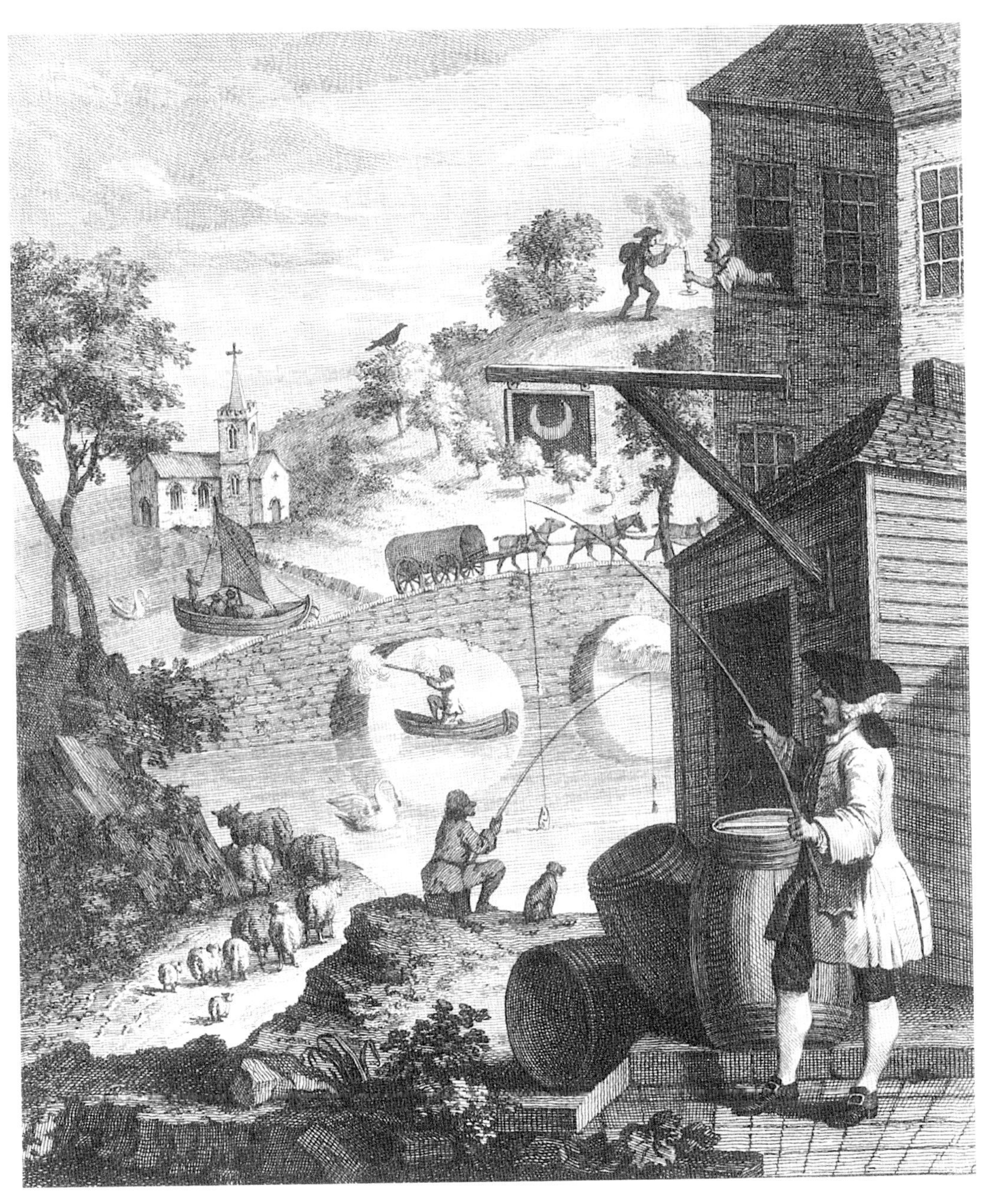

FIG. 2.12. William Hogarth (Luke Sullivan, engraver), *Satire on False Perspective*. From John Joshua Kirby, *Dr. Brook Taylor's Method of Perspective Made Easy* (London, 1754). Getty Research Institute, Los Angeles.

and two tented leaves. Bartram stitched these parcels together with the stippled lotus stems and the flowering plant on the drawing's far left edge, creating an interlocking composition. Formal rhymes enhance the effect—the repeated use of an ovoid shape links the body of the great blue heron, the lotus bud, and the thinly inscribed vein patterns on the lotus leaves.[59] The drawing's division by thirds and fifths, along with the unity of its composition, faithfully adheres to Hogarth's instructions.

In this exceptionally literal interpretation of the picturesque, Bartram substituted elegant design for spatial coherence. The rejection of a cogent perspectival scheme seems to follow the model of Mark Catesby, who suggested that the flatness of his renderings yielded greater representational transparency. Yet Catesby's images are flat through and through, whereas Bartram's drawing of the lotus exhibits unanticipated shifts between surface and depth. These shifts emerge from his refusal to integrate the drawing's different representational conventions into a plausible portrayal of three-dimensional space. Instead, the drawing's seams lie bare, offering a visually pleasing but bewildering vista. Just as Bartram explored the American wilderness to pursue unfamiliar species for his European colleagues, so too must the drawing's viewers explore and assess this unconventional landscape. Bartram even reinforced the theme of pursuit in his alteration of the heron, which hunts its prey with greater alacrity than in Edwards's etching, and in the inclusion of the Venus flytrap, which patiently attends the dragonfly.

The Pleasures of Exploration and Pursuit

If the tenets outlined in Hogarth's *Analysis* offered Bartram a template for the picturesque, they may also have offered a model for conveying a sense of exploration and pursuit. Hogarth's aesthetic treatise was enormously popular in British North America, where the absence of art academies required aspiring artists to train abroad or to teach themselves through manuals. Hogarth's prolific print production made his artwork well known in the colonies, and both his prints and his treatise were regularly advertised for sale at stationers' shops in Philadelphia. Indeed, the Library Company voted in 1764 to acquire the full suite of Hogarth's prints, which were finally purchased and delivered to Philadelphia in 1767.[60] They acquired their first copy of the *Analysis* between 1757 and 1765 and a second copy when they merged with the Union Library in 1769.[61]

American colonials may have acquired the *Analysis* because of Hogarth's celebrity, but they no doubt admired it for its argument that beauty is accessible and understandable to all. Hogarth situated his text in direct opposition to the gentleman connoisseurs, "nature-menders," and theorists who wished to keep beauty the rarefied province of the elite.[62] The *Analysis* was conceived, in part, as a rebuke to the aesthetic theory of Anthony Ashley-Cooper, the third earl of Shaftesbury, who maintained that the observer of beauty must be disinterested and distanced from what he (invariably "he") perceives.[63] John Barrell details the social and political implications of a premise that limited taste to the aristocratic gentleman, since those who labored among the things of this world had neither the leisure nor the distance to be disinterested.[64] Shaftesbury's theory promoted a geometricized ideal of beauty that required familiarity with classical references and a mastery of arcane rules of proportion. But according to Hogarth's *Analysis*, beauty could be found anywhere, from the finest Michelangelo to the humblest ash leaf, and by anyone. Where Shaftesbury celebrated the idealized models of classical antiquity, Hogarth proposed an aesthetic theory that was both contingent and dynamic, tied to nature's flux. For colonials who were far from the artistic centers of London, Paris, and Rome, Hogarth's must have been a welcome view.[65]

The *Analysis* presents "fitness" as the overarching requirement for beauty, since a beautiful form is one suited to its object and function. Fitness establishes a proper balance of variety and uniformity, simplicity and intricacy, and yet—because this balance is not a given—Hogarth's theory also necessitates a process of discovery. His text often reads as a paean to pursuit: "Pursuing is the business of our lives, and even abstracted from any other view, gives pleasure. Every arising difficulty . . . gives a sort of spring to the mind, enhances the pleasure, and makes what would else be toil and labour, become sport and recreation." According to the *Analysis*, aesthetic pleasure is inseparable from this act of discovery. "Wherein would consist the joys of hunting, shooting, fishing, and many other favourite diversions, without the frequent turns and difficulties, and disappointments, that are daily met with in the pursuit?" Forms that keep the eye in motion are enjoyable, even beautiful, just as a cunning old hare pleases the hunting dog more readily than the prey easily caught.[66]

As Frédéric Ogée has made clear, Hogarth illustrated the pleasures of pursuit not only through verbal metaphors but also through the visual associations found in the treatise's explanatory plates.[67] Each plate features a central

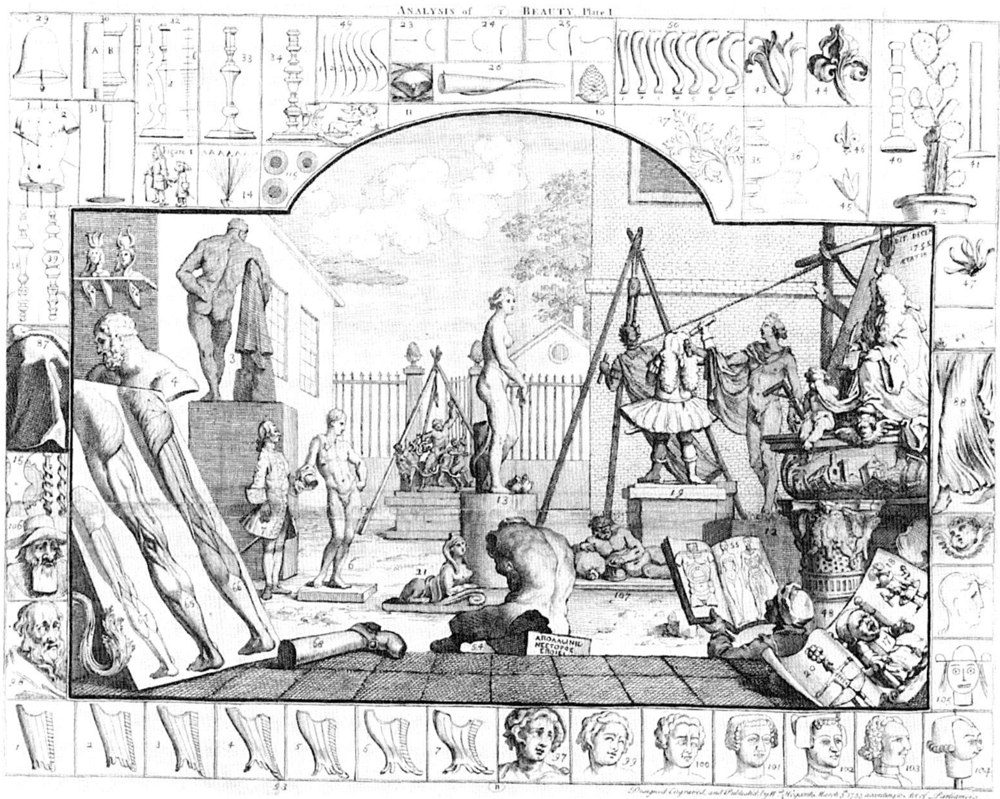

FIG. 2.13. William Hogarth, *Plate I*. From *The Analysis of Beauty* (London, 1753). Paul Mellon Collection, Yale Center for British Art.

scene—a sculpture yard in the first, a dancing party in the second—with a series of numbered figures ranged round its border (figs. 2.13 and 2.14). Despite the taxonomic effect of the numbered figures and gridded border, the represented objects seem almost arbitrary. The perimeter of plate 1 features, among other things, cross sections of molding and both natural and stylized plants. Its central scene, though carefully organized according to linear perspective, hosts an equally unexpected assortment of figures. While it portrays patrons examining casts of classical statuary, as befits a sculpture yard, it also depicts such unanticipated vignettes as a turbaned man reading an anatomy text. The relationships among the scene's objects are associative, and Hogarth encouraged viewers to pursue these associations and patterns.

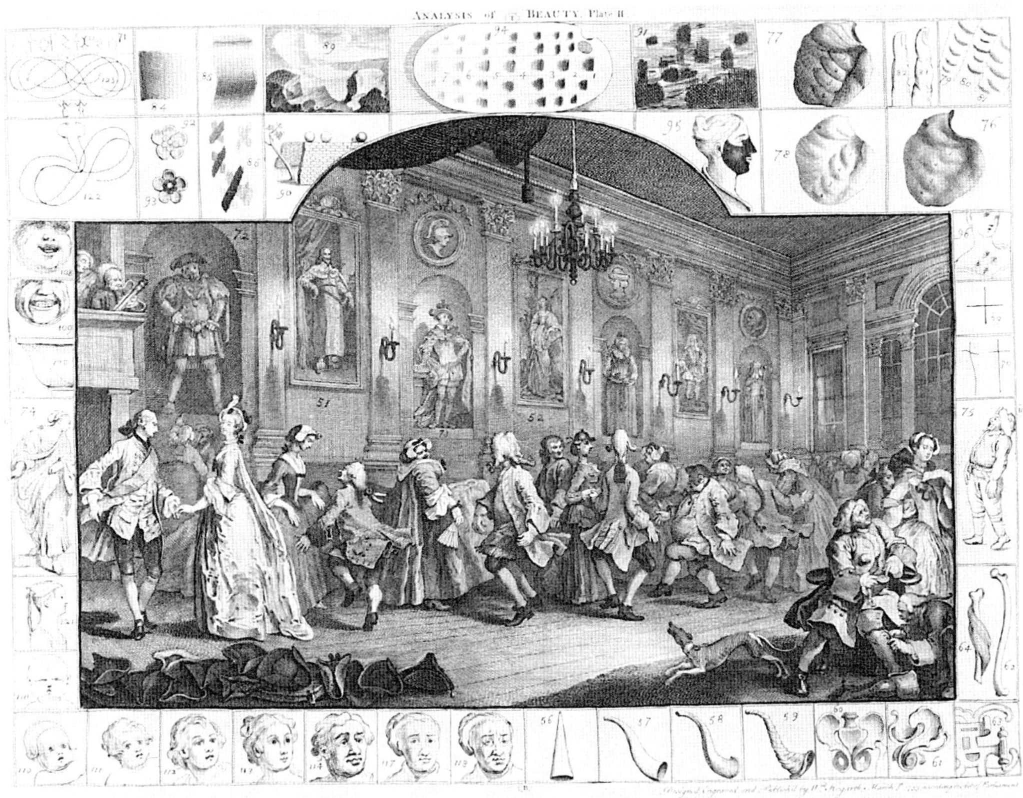

FIG. 2.14. William Hogarth, *Plate II*. From *The Analysis of Beauty* (London, 1753). Paul Mellon Collection, Yale Center for British Art.

Hogarth's description of "pleasing forms" offers just one example of this pursuit. His text notes how the curves of the molding in figures 35 and 36 of plate 1 are divided into "well-shaped quantities" that give the molding a pleasing effect, not unlike the parsley leaf in figure 37, which is similarly divided into "three distinct passages."[68] While the text identifies the figures, it is the reader's task to find and compare them. Figure 37 is to the immediate left of figures 35 and 36 in the upper right quadrant of the picture, making them relatively easy to locate, but other references require a concerted search. For instance, when the text cites figure 48, a capital composed of hats and wigs that exhibits the same balance of simplicity and variety as the molding and parsley leaf, it would seem that it, too, should be located in the upper right quadrant of the plate.[69]

Yet the area shows only figures 35–37, 40–47, and 50, forcing the eye to scan the frame and enter the perspectival space before locating the capital next to the turbaned man.

Ogée refers to this process as an act of "dynamic perception," in which the viewer must actively explore and make sense of the scene, and he likens it to Hooke's sequence of observations through the microscope.[70] Despite the equivalence Ogée sees between Hooke and Hogarth, however, their resulting illustrations are quite different. Hooke engaged in a time-consuming method of close examination and synthesis in order to discover an object's form, yet he did not convey this method visually, instead opting for a seamless composite that concealed the steps between observation and engraving. Hogarth's plate, on the other hand, maps out those steps. The eye must move through the image, observing and relating seemingly incompatible figures, in order to grasp his argument. "Hooke defined observation and analysis as a circulatory process of constant exchanges," Ogée notes, yet it is not in Hooke but in Hogarth that this process is given visual expression.[71] In his *Analysis*, Hogarth presented his theory of aesthetics as well as the means by which he developed that theory.

And it is this element, Hogarth's method of laying out how he came to know the world, that would likely have been of greatest interest to Bartram. As Ogée and many other scholars have made clear, the *Analysis* was no mere painter's manual; the book originated in the coffeehouses and studio spaces of St. Martin's Lane in London, where physicians, anatomists, and botanists rubbed shoulders with painters, sculptors, and designers. The various threads of Hogarth's social and intellectual milieu course through his treatise and help to explain its broad and diverse audience.[72] As much as Hogarth's focus is on taste, he excavated much larger questions about how to see, navigate, and represent a rich and messy world full of interests, intrigues, and bewildering details. How should one portray life's fecund variety and vibrancy? In the *Analysis* Hogarth hazarded an answer.

The plates for the *Analysis* call into question the naturalness of rationalized modes of representation, such as the specimen drawing and the rigid perspectival scheme. For Hogarth, these modes of representation are unequal to portraying the astonishing diversity of the world. Although the plates' gridded frames read as a series of Hooke- or Ehret-designed illustrations, each featuring an object or a set of objects isolated against a blank background, Hogarth juxtaposed them against the one-point perspective of the central scene, creating a jarring union of different representational modes. In tracking the formal comparisons he offered in his text, the eye shifts between surface and depth as it moves between taxonomic and perspectival grids, indicating that neither is sufficient to capturing the variety and mutability of nature. The nearly contemporaneous frontispiece for *Dr. Brook Taylor's Perspective Made Easy* starts to seem less satirical when considered in relation to the *Analysis*; or, rather, it appears to satirize linear perspective as much as or more than its false application. Like the plates of his treatise, Hogarth's frontispiece rejects the concept of representational transparency and, by toggling between flatness and fullness, demands the viewer's active engagement.

Bartram's rendering of the lotus equally triggers such participation. He constructed a bewildering sense of space to keep the eye motile, scanning the composition in an attempt to makes sense of its shifting scales and points of view. Through this act of pursuit, the viewer encounters the objects and concepts by which Bartram came to know the natural world, signaled by the specimenlike leaf, the Linnaean flowers, and Edwards's heron. Moving from his material observations to their rational processing, he indicated that natural knowledge is neither self-evident nor the product of a distant and disinterested curiosity but, rather, is the result of one's own physical and intellectual investment in living nature. Eschewing the notion of Hooke's true form, Bartram mapped out his different stations of pursuit, sensory observation, and intellectual processing in his natural history. The drawing thus serves more as a field guide for the viewer than as a transparent window onto North American nature.

When Collinson received Bartram's drawing he was transfixed. "I and my Son opened my Ingenious Fr^d Williams, Inimitable Picture of the Colocatia," Collinson wrote in February 1768. "So great was the Deception it being a Candle Light that we Disputed for Some Time weather it was an Engraveing or a Drawing it is really a Noble peice of Pencil Work. . . . I will not Say more in commendation because I shall Say to Little where So Much Due."[73] With the drawing's curious spatial construction, visual citations, and introduction of nondescript species, Bartram created an image that was as engaging and complex as nature itself. In the image he marked not just what he had come to know and understand about this plant species that was so elusive to European virtuosi; he recorded the very process by which he had come to know it. One can only imagine Collinson and his son hunched in the dim winter light, both delighted and perplexed by Bartram's wholly unexpected—and, indeed, inimitable—depiction of natural knowledge.

Chapter 3

Vital Matters

Redefining Natural Knowledge

We might read William Bartram's portrayal of the American lotus as a manifesto of sorts, an articulation of his own theory of representation. Peter Collinson quickly put the drawing to work in hopes of securing patronage on his behalf. Happily, in July 1768, Collinson could inform the Bartram family of a meeting with John Fothergill that concluded with a commission: "This Morning Doc Fothergill came and Breakfasted Here—as I am always thoughtful How to make Billeys Ingenuity turn to Some Advantage—I bethought Mee of Showing the Doc^r—His Last Elegant performances—He Deservedly Admired them & thinks So fine a Pensil Is worthy of Encouragement."[1]

After Collinson's death in August 1768, Fothergill became William Bartram's most important patron. William began to make drawings for him in the 1760s, and in turn Fothergill later underwrote his botanizing trip to the southern colonies in the 1770s. In an undated letter, Fothergill assured John Bartram of his support of his son: "For his sake as well as thine, I should be glad to assist him. I Ie draws neatly, has a strong relish for natural History and it is a pity that such a genius should sink under distress," gently referring to William's previous challenges.[2] His letters offer clear instructions for these drawings, requesting they be "exact in the parts of fructification; and where these parts are very diminutive, to have them drawn a little magnified."[3] Fothergill desired proper Linnaean portrayals of specimens, so that he might classify them.

Much like the depiction of the American lotus, Bartram's drawings for Fothergill both subvert and resist the conventions of eighteenth-century natural history representation in order to map out his empirical experiences and construction of knowledge. At the same time, they indicate the limits of that knowledge—of what the philosopher or naturalist could reasonably know. In the second half of the eighteenth century, the prevalence of Linnaeus's taxonomic system notwithstanding, many observers found nature's order increasingly impenetrable, in part because of a resurgent interest in sentient matter. In his *Essay Concerning Human Understanding*, John Locke suggested that all organic matter may have the capacity to think and perceive, a proposition that gained credence in the mid-eighteenth century following the popularization of microscopy and new investigations into the freshwater polyp or hydra (*Hydra vulgaris*). The possibility of sentient matter challenged a number of seemingly self-evident distinctions. If matter was to some degree self-organizing and responsive to stimuli, then the boundaries between species, genera, or even between plants and animals could hardly be rigid or stable. Such a development threatened the hierarchy of the Great Chain of Being or *scala naturae* that ordered all organisms according to their inherent complexity.[4]

Fothergill sought verbal and visual descriptions that would firmly articulate taxonomic boundaries and reinforce this hierarchy, but Bartram was less committed to such a project. His skepticism regarding taxonomic systems, and whether a natural order might ever be accessible to human understanding, shaped his approach to natural history. He was taken with nature's wonder and beauty but also by its instability and unruliness: its entwining of life and death, violence and mercy, made him hesitant about making declarative statements regarding its structure. In his drawings he instead proposed a wholly different kind of knowing, one that embraced the disruptiveness of lived experience and nature's metamorphic processes.

Microscopy's Magazine of Wonders

In August 1744, the pages of the *Pennsylvania Gazette* announced the recent arrival from London of a solar microscope, whose wonders would be demonstrated at Mr. Stephen Videll's Philadelphia schoolhouse, located on the corner of Second and Chestnut Streets. This newly developed device combined the magnification powers of a double microscope with the projection capabilities of a camera obscura, allowing groups of people to observe the microscopic object simultaneously. For eighteen pence, the advertisement promised, attendees could see the spectacle of objects magnified "to the most surprising Degree." They could observe the miniature animalcules in fluid and paste and witness the pulsing heart and viscera of a frog, fish, flea, and louse.[5] In mid-eighteenth-century Philadelphia, microscopy had moved from highly specialized research into public spectacle, at once both entertaining and edifying.

The popularity of the microscope was marked not only by public displays but also by the acquisition of apparatus and published texts, ordered directly from abroad or purchased from Philadelphia's London Book-Store.[6] They could also be accessed at learned institutions such as the Library Company of Philadelphia, which owned a double microscope, as well as copies of Henry Baker's *The Microscope Made Easy* (1742), his *Micrographia restaurata* (1745), and George Adams's *Micrographia illustrata* (1746), among others.[7] Baker, a celebrated naturalist and publisher, and Adams, a London instrument maker, helped popularize the practice of microscopy among nonspecialists in the second half of the eighteenth century. The experiments described in their publications—which cribbed from texts by such eminent seventeenth-century microscopists as Hooke, Jan Swammerdam, and Antonie van Leuwenhoek, as well as from the published experiments of their contemporaries—were intended to introduce the public to the complexity of the living world. Baker observed that magnification transformed "the whole Universe into a Magazine of Wonders." Adams thus summarized its effects: "The whole Earth is full of Life; there not being a single Tree, Plant, or Flower but that affords Food and Shelter to a Species of Inhabitants peculiar to itself. And then if we call in the Assistance of Art, what a new Scene of Wonder opens to our View? What an infinite Variety of living creatures present themselves to our Sight?"[8]

It is unknown whether William Bartram attended any public "philosophical lectures" that featured optical devices and discoveries, though we can assume he engaged in microscopical study. His father had conducted microscopical observations of plants as early as 1739 and had also been gifted a copy of Baker's *Microscope Made Easy* in 1758. Baker's book in particular seems to have had an enormous influence on William, who quoted it either in paraphrase or verbatim throughout his writings.[9] His natural history often evokes the experiential effects of this optical practice. His verbal and visual descriptions present a disorienting yet wondrous natural world, one in which life and death entwine and perpetually shifting perspectives reveal nature's manifold complexities. Only an accretion of many different perspectives—a magazine of wonders—may offer a sense of nature's intricacy, even though it may never fully delineate it.

Although the visual rhetoric in many microscopy manuals suggested a perfectly self-evident natural world, the underlying premise was that nothing is as it first appears. Something as simple as, say, the rot on fruit, bread, and leather radically transforms when magnified.[10] Both Baker and Adams, recapitulating Hooke, described how a spot of mold on a sheepskin binding transforms via the microscope into a landscape covered with tender, tiny mushrooms, each growing from a different pore in the leather. The accompanying illustrations—again, copied directly from Hooke—reinforce the idea of a miniature vista, depicting the mold's long, thin stems, spherical knobs, and irregular blooms (fig. 3.1, top). The terrain from which these delicate plants emerge appears to undulate gently, dotted with pebbles and undergrowth. When viewed through the lens, a site of decay becomes, instead, a landscape brimming with vigorous life.[11] Even as the illustrations adopt a rhetoric of confident self-evidence, the texts themselves reveal the bewildering nature of the microscopic object.

Microscopy's subversion of self-evidence stemmed from the technology's unhinging of scale and perspective. Illustrations of the magnified blight on rose leaves present it as though it were a rolling field flecked with seedpods (fig. 3.1, bottom), yet the accompanying verbal descriptions explain that these pods are less than 1/500th of an inch, leading to a vertiginous telescoping of scale. In *Micrographia restaurata*, Baker adopted Hooke's own incredulous language, querying, "If these Pods, as is highly probable, contain Seeds, and the Size of those Seeds bear such a Proportion to that of the Pod, as we find between the Seeds and Seed Vessels of *Pinks, Columbines, Poppies,* &c. how inconceivably minute must each of those Seeds be?"[12] In *Micrographia illustrata,* Adams

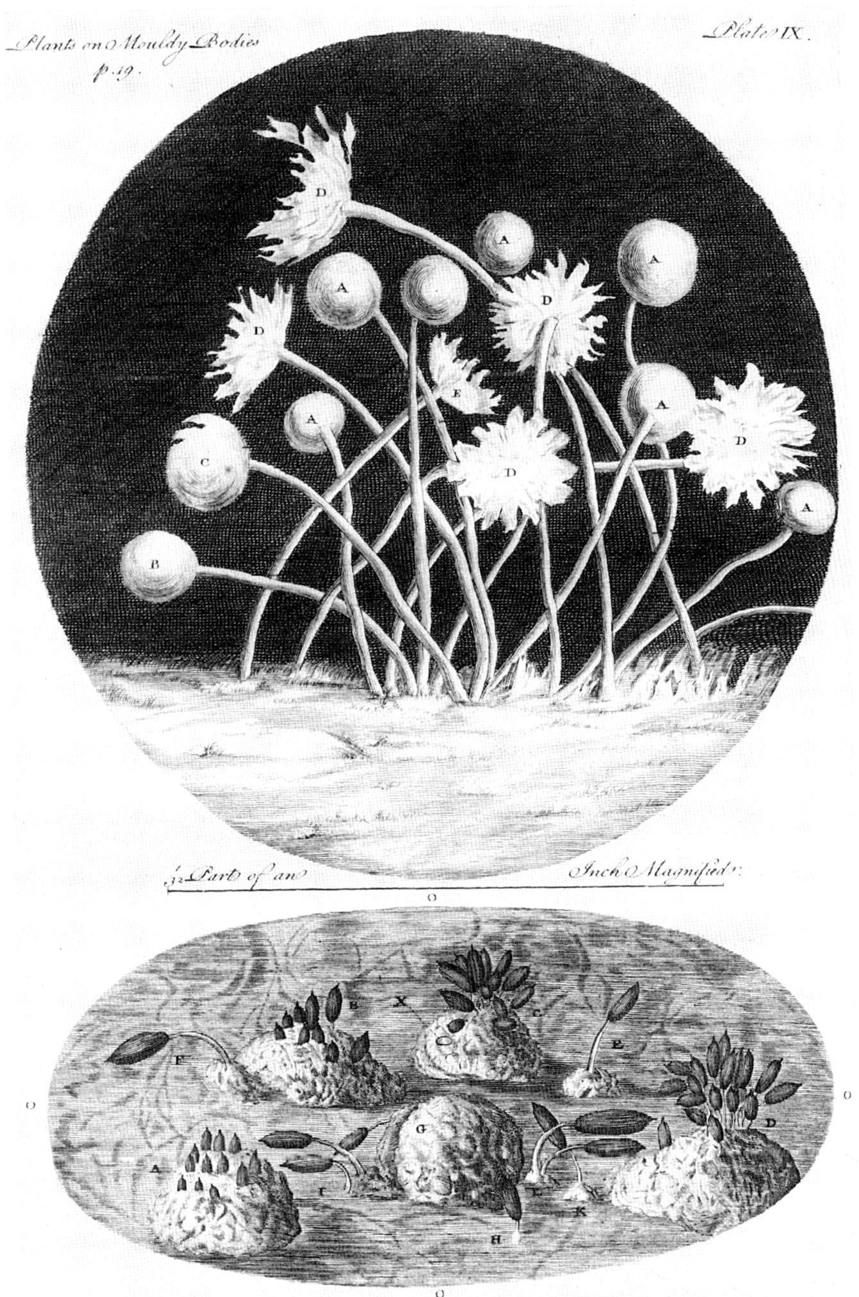

FIG. 3.1. Henry Baker (after Robert Hooke), *Plate IX*. From *Micrographia restaurata* (London, 1745). Bruce Peel Special Collections, University of Alberta Library.

likewise expressed awe over such shifts in scale by comparing the blighted leaves to mosses and mahogany. Blight, he pointed out, is but one 1/1,000th the bulk of moss, which in turn is only 1/2,985,984,000,000th the size of the largest West Indian tree. Adams's comparison was intended to serve as a paean to the incredible intelligence and power of the Creator, yet the juxtaposition also highlights the disorientation caused by the relativity of scale.[13] Nested in each subvisible surface are more and infinitesimally smaller forms of life awaiting discovery.

Bartram's evocation of new worlds, his shifts between the massive and the minute, and his delineation of mutable forms all draw on the bewilderment caused by the magnified view. In *Travels*, for instance, nothing is ever as it seems. Floating islands of water lettuce along the St. Johns River appeared abundant with wildlife and "old weather-beaten trees, hoary and barbed," just as Bartram's canoe diminished "to a nut-shell, on the swelling seas," while he traversed Lake George during a storm.[14] Moreover, many of the reptiles he saw in the lake's waters were not reptiles at all, but ahingas or snakebirds. "If this bird had been an inhabitant of the Tiber in Ovid's days," Bartram observed, "it would have furnished him with a subject, for some beautiful and entertaining metamorphoses."[15] The transformations he described in the text are often born of changes in perspective or of the difference between objects up close and far away.

These perspectival changes are intentional, part of Bartram's investigation of the complexity of the natural order. One of the first anecdotes in *Travels* begins with him standing high atop a hill along the Florida coast, taking in a broad vista that encompasses the Atlantic Ocean, the Mosquito River, and aromatic groves of orange trees and sweet myrtle. He descended the hill to collect botanical specimens from the shrubs, but as he approached, he narrowed his focus to a spider on one of the leaves; he described its buff color, singular markings, and its predatory behavior. The spider cautiously hunted a bumblebee, which it soon trapped and fatally injured.[16] The text transitions from a panoramic view, encompassing at least a quarter of a mile, to a dramatic scene of life and death played out on a single leaf, evoking microscopy's shifts between the expansive and the circumscribed.

In this passage, each vantage point offers something different: the first yields a prospect and the second an ecological interchange nested deep within that prospect. These spatial shifts derive from period microscopy texts, as do Bartram's reflections on moral economy and the alignments between human and animal. On the preceding pages, Bartram offered an especially affecting narrative, in which he and his traveling companion made stealthy, "oblique approaches" on a mother bear they were hunting.[17] After taking it by surprise and killing it, Bartram noted that the bear's cub pawed its lifeless body, appearing to weep and mourn. Bartram was so struck with compassion for the bear and its offspring that he began to reflect on his role as an accomplice to this "cruel murder" and on the morality of the natural world.[18] In describing the spider and the bumblebee, he observed how predator likewise approached prey "with slow steps obliquely, or under cover of dense foliage," much as he and his traveling companion tracked the bear.[19] Bartram's description of the spider's pursuit may be drawn from his own observations, but it also aligns with Baker's discussion, in *Micrographia restaurata*, of the hunting spider's subtle craftiness in tracking its prey.[20] Ultimately the juxtaposition of these two acts of violence inspired Bartram to ask whether such behavior can be reprehensible in one context while excusable in the other. He landed on no definitive conclusion but instead rested uneasily with the recognition that predator and prey, life and death, are part of nature's design, decreed by God but no less unsettling for it.

The relationship between hunter and hunted is a recurrent concern in Bartram's writings and drawings but is perhaps most notable in his descriptions of the Alachua Savanna in East Florida, now known as Payne's Prairie. The site has long been considered one of the region's more notable natural curiosities. The reservoir along its northern border, the Alachua Sink, is a product of the Florida peninsula's water-conducting limestone, which can dissolve over time and create sinkholes and underground springs. The sink serves as a conduit between the surface water of the savanna and the groundwater of the aquifer. Depending on rainfall, the site alternates between boggy marsh and lush grassland. In the seventeenth and eighteenth centuries, this grassland provided the Spanish and the Seminole Indians pasture for their horses and cattle; in the late nineteenth century, obstruction of the sink transformed the savanna into a lake, and for nearly twenty years it could be crossed only by boat.[21]

The Alachua Savanna's topographical instability made it an especially compelling site for Bartram to explore. In perhaps the most oft-quoted passage of the *Travels*, he registered his first sight of it with an exclamation of wonder that recalls the language of Baker and Adams: "The attention is quickly drawn off, and wholly engaged in the contemplation of the unlimited, varied, and truly astonishing native wild scenes of landscape and perspective. . . . How is the

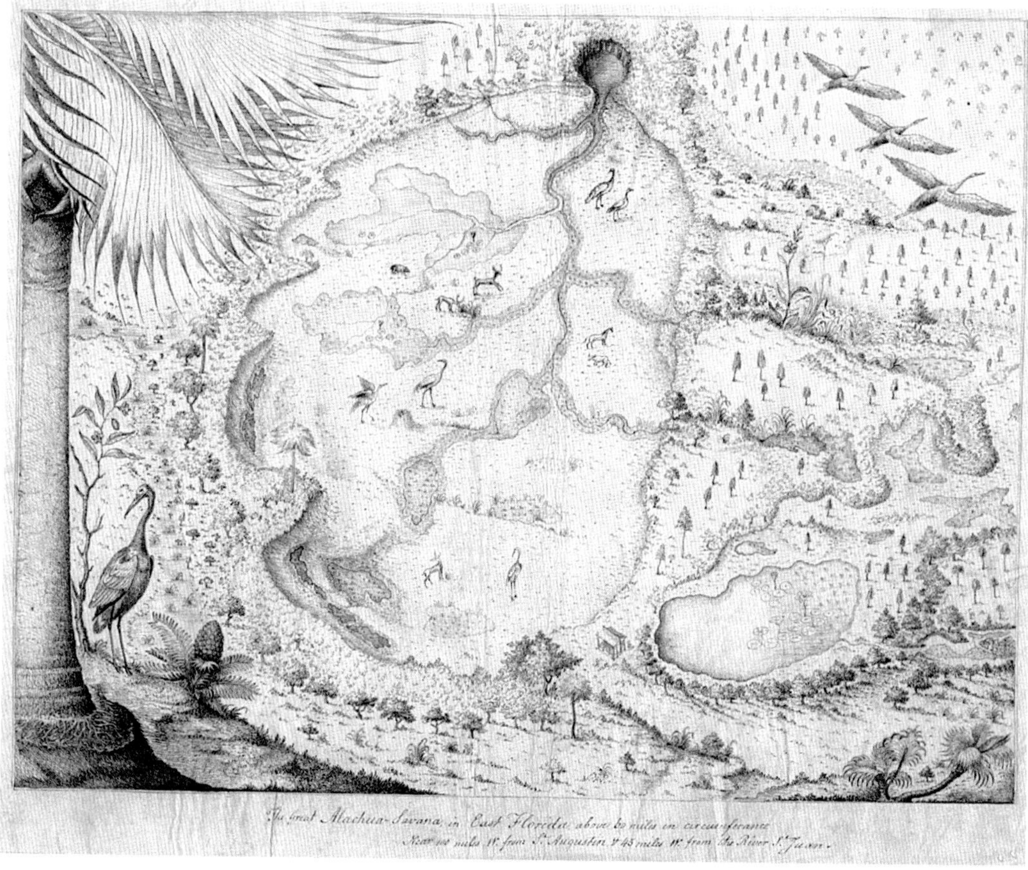

FIG. 3.2. William Bartram, *The Great Alachua-Savana, in East Florida*, n.d. Courtesy of the American Philosophical Society.

mind agitated and bewildered, at being thus, as it were, placed on the borders of a new world!"[22] His rhapsodic effusions over the soaring sandhill cranes, distant waving lotus blossoms, and groves of trees capture the same sense of the marvelous as the discovery of minute organisms. But he did not end his account of the savanna with this declaration of a new world; instead, he pursued various perspectives, moving in a meandering path that allowed for broad vistas interspersed with contracted views. He described with equal attention the glowing sunset, the green coves that "scollop" the edges of savanna, and the thin fragile snakes at his feet, "innocent and harmless as worms."[23] These changes

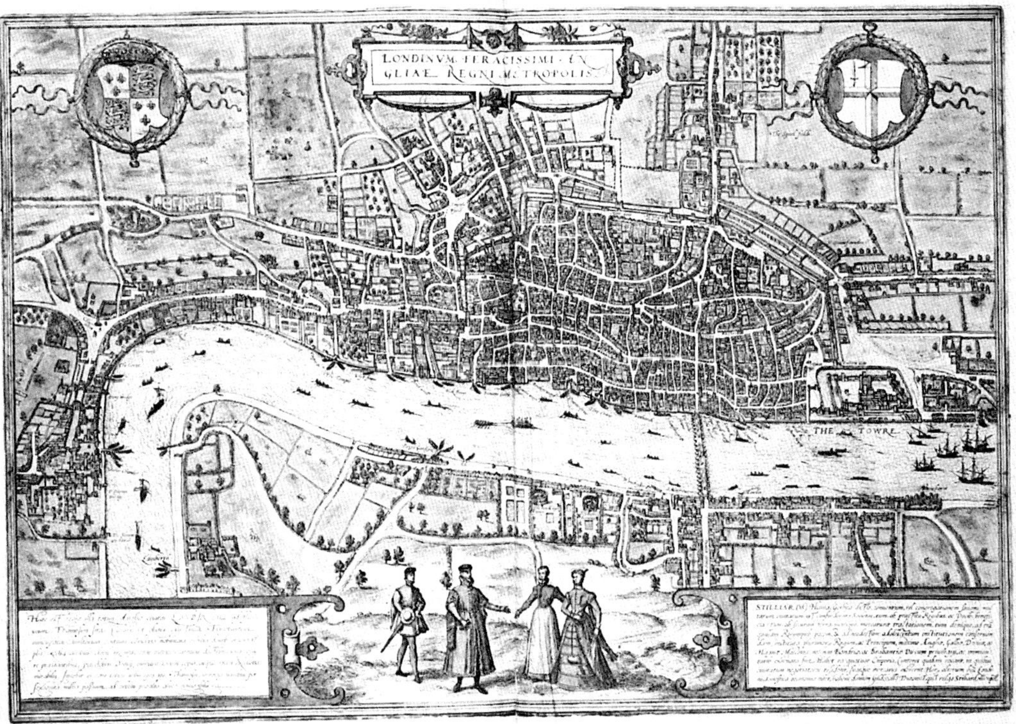

FIG. 3.3. *London*. From Georg Braun and Frans Hogenberg, *Civitates orbis terrarum*, vol. 1 (Cologne, 1572). © The British Library Board.

in perspective reveal that the savanna offered far more than the initial prospect suggested, as the very topography shifted beneath him. Waving fields become tarns thick with sedge, and solid earth yields to swampy morass.

This instability transforms the savanna into a site of both fecundity and fatality. The waters that flow from the sink provide a habitat for numerous fish, while also nourishing the reeds and grasses. As the rivulets evaporate, however, the fish frantically crush into the sink, where they become easy prey for alligators. Nature is structured in such a way that predation is inevitable. Those fish that do not make it back to the sink become trapped on land and perish, where their rotting bodies soon overpower the scent of blooming trees.[24] The savanna, like magnified bindings and blighted leaves, reveals the interlacing of life and death.

The Alachua Savanna also underscores the overlaps between animal and human, which Bartram highlighted in his visual rendering of the site. He originally made two drawings of it, one that he retained and one that he submitted to Fothergill along with a written report in 1775. *The Great Alachua-Savana, in East Florida* (fig. 3.2)—the version Bartram kept for himself—is not a fully aerial, abstracted view but instead borrows the high horizon of many sixteenth- and seventeenth-century maps, made popular in Georg Braun and Frans Hogenberg's six-volume *Civitates orbis terrarum* (1572–1617; fig. 3.3). These volumes formed a compendium of city and town views from around the world, and the illustrations emphasized key topographical and architectural features. According to Lucia Nuti, the intention was to develop maps rooted in direct observation rather than from abstract geometry; the views provided topographical knowledge born of personal experience.[25]

In *Civitates orbis terrarum* the mapmakers' desire to convey both a total knowledge of a site and the subjective origin of that knowledge places the viewer in a liminal position. Svetlana Alpers identifies such works as midpoints between cartography and landscape representation, which leave the observer both separate from and embedded in the view.[26] In his drawing of the Alachua Savanna, Bartram embraced this tension by including in the lower left corner a savanna crane beneath a tall columnar tree. Placed firmly in the foreground, these serve as our point of entry into the scene. The viewer, however, is soon spatially untethered, as no middle distance bridges the foreground and the aerial flatness of the savanna. The inconsistent scale of the flora and fauna that dot the drawing's center reinforces the lack of groundedness and stability—cranes are nearly twice the size of deer and horses, and overgrown marsh plants tower over trees. The spatial incongruities of this drawing produce so many competing viewpoints that, Thomas Hallock observes, "the search for perspective becomes its own theme."[27]

The prominence of the crane and the stylized appearance of the tree— seemingly more architectural feature than natural production—also works to collapse distinctions between intellectual construct and natural world. The bird's position at the point of entry makes it read as a stand-in for Bartram himself; Nuti observes that the makers of such maps often included themselves in the foreground, surveying the landscape, to reinforce the authenticity of these experiential views.[28] In *Travels* the alignment of human and animal—first addressed in his juxtaposition of the bear hunter and the stealthy spider—occurs repeatedly. Bartram again and again described nature in humanly crafted terms: fields are enameled with flowers, cormorants compose themselves into pictures, and cranes form disciplined squadrons.[29] Bartram similarly aligned nature and culture in this rendering of the Alachua Savanna by using a representational model reserved for city views to portray a natural wonder.

In aligning nature and culture in this way, Bartram was likely influenced by Baker's *Microscope Made Easy*, whose penultimate chapter compares works of art with works of nature. Baker likened the point of a needle to the sting of a bee and Brussels lace to a silkworm's web. In both instances, the cultural production pales in comparison to its natural counterpart; under magnification it is blemished and uneven, while the natural production displays an unexpected elegance of construction. Yet Baker still employed cultural terms to define the richness of the natural world: insects are set with polished jewels, while the organization of animalcules reveals exceptionally fine workmanship.[30] Both Bartram and Baker deployed such comparisons to highlight the wisdom of the Creator, but their juxtapositions also challenge a hierarchical Chain of Being. In a draft essay comparing human and animal nature, Bartram (adapting Baker's own comparisons, and even his language) observed, "I will agree that it is impossible that any Animal will or can weave a piece of Brocade[,] Make . . . A Sewing needle no larger [than] the Eye lash & small as a hair which shall contain Eleven others, one within another, & It is equally imposseble, For Man to make a Spiders Web [or] A Honeycomb with Wax & Honey." The beauty and elegance of the meanest natural productions provides, in Baker's words, "convincing Proofs of their not being so insignificant as we presumptuously suppose they are."[31]

Where Bartram's *Great Alachua-Savana, in East Florida* upends a sense of rational order, the drawing he submitted to Fothergill seems at first to reinforce it, borrowing the tools of mathematical measurement and order commonly found in microscopy illustrations. Similar to the plate depicting feathers in *Micrographia restaurata*, Bartram's *View of Alatchua Savanah* (fig. 3.4) is anchored along its bottom border by a graphic scale. Likewise, letters key different elements of the image to textual identifications that include the Seminole town of "Cuscoela" (Cuscowilla) and a neighboring lake, the Alachua Sink, a shuttered trading post, and an east–west road. The scale and key echo the rationalizing devices found in scientific illustrations and topographical maps and are intended to connect the site's abstracted representation to its physical existence.

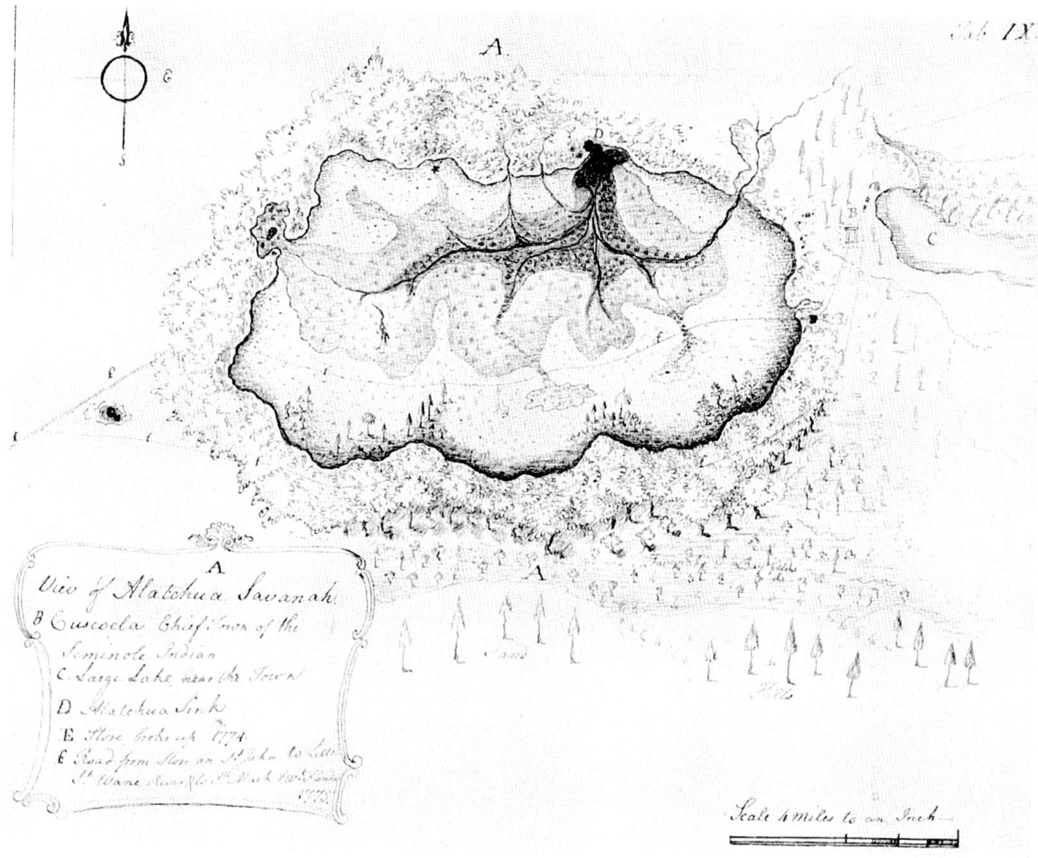

Fig. 3.4. William Bartram, *View of Alatchua Savanah*, 1775. © Natural History Museum, London.

Bartram's use of legible symbols to depict the savanna's surface also reinforces a sense of stability, as small stylized pines and bubbly, scrubby marks designate evergreen and deciduous trees and repeated clusters of three short lines signify grassland. Variations in tone indicate different degrees of moisture in the soil, ranging from the white of the grasslands to the grey patches of marsh, to the dark snaking rivulets that flow to and from the sink. As Francis Harper points out, however, the map is curiously askew. With the town of Cuscowilla to the east of the savanna rather than the south, Bartram transformed an ostensibly coherent and stable picture into a perplexing one.[32] At each turn in his natural history, Bartram seemed to have been in thrall to the unsettling effects of microscopy, which he conveyed in his drawings. His juxtaposition of vantage

94

points, disruption of scale, and challenge to conceptual categories all recall the bewilderment and wonder of the magnified view.

A Disruptive Vital Principle

In the seventeenth and eighteenth centuries, the microscope not only offered up new worlds to its practitioners but also influenced new understandings of mind and of its relationship to the living organism. The dualism inherent in Descartes's mechanistic model of the universe held that mind and body were distinct entities, one immaterial yet active, governing perception, intentionality, and emotional states, and the other material yet inert, possessing only physical properties such as form, weight, and extension. Microscopical studies, however, suggested that the division between the two was hardly so complete.[33] The English physician Francis Glisson, for example, proposed the notion of a "natural perception" in matter in his *De natura substantiae energetica* (1672). Although distinct in some ways from sensibility—that is, perception attained through the sense organs—natural perception was a difference of complexity and organization rather than of kind.[34] Even Robert Hooke, considered one of the preeminent mechanists of his era, was hardly doctrinaire about the distinction between mind and body.[35] In the *Micrographia* he often presented the world as a machine of inert components, as in his description of the louse's heart as "a contrivance, somewhat resembling a Pump, [or] pair of Bellows" that moved blood and other fluids through its body. Other parts of the text seem to counter this view, however. Hooke's belief that some insects may spontaneously generate from "Vinegar, Meal, musty Cakes, &c." suggested the possibility of a vital principle inherent in matter.[36]

For some naturalists and philosophers, a vital principle was the most apt response to questions that seemed otherwise impossible to answer.[37] A mechanistic worldview could not account for the heat of living animal bodies, nor could it account for the motion of muscle fibers when disconnected from the central nervous system. In the mid-eighteenth century the Swiss anatomist Albrecht von Haller, curious about these involuntary movements, engaged in a series of gruesome experiments to assess their origin. By "irritating" different parts of vivisected animals through cutting and burning, Haller observed that muscle fibers could react to stimulation without the intervention of the central nervous system. He called this property of self-movement irritability, which he contrasted with sensibility, much like Glisson. Yet Haller maintained a strict division between irritability and sensibility and did not suggest a possible continuum between the two, as Glisson had in his study of natural perception. Such a division preserved the dualism inherent in Cartesian thought by keeping any vital property in matter separate from the operations of mind.[38] Yet other experiments during the period, such as Abraham Trembley's microscopical studies of the freshwater polyp, breached this boundary by suggesting that intentional movement might be intrinsic to organic matter. Charles T. Wolfe has described this philosophical position as vital materialism.

Trembley was a Swiss naturalist who began his career as a tutor to the sons of William Bentinck, a member of the Dutch and English nobility. While searching for insects with his pupils at the Bentinck estate in the Netherlands, he came across a minute plantlike animal or animal-like plant, which he described in his correspondence with René-Antoine Ferchault de Réaumur. In his initial letter, dated December 15, 1740, Trembley observed that the organism possessed a tubular body that bore no internal or external parts, save for long appendages or "threads" that it was capable of moving at will. While its exceedingly simple anatomy suggested plant life, its capacity for spontaneous motion—it could extend and contract, thus propelling itself by flipping end over end—indicated an animal nature. Still, Trembley wrote, "there is much resemblance between what happens to this animal, and plants. . . . Perhaps the threads . . . are sorts of roots and that what I call an animal is perhaps an ambulant and sensitive plant, but which has characteristics very different from all those which one knows."[39] Trembley was unaware that the creature had already been discovered by Van Leeuwenhoek, who tentatively described it as an animalculum, as did the French naturalist Bernard de Jussieu. Of the three, however, only Trembley engaged in a rigorous process of microscopical observation and experimentation to determine the polyp's status.[40]

The three most important criteria that Trembley adopted for distinguishing between plants and animals—movement, reproduction, and nutrition—came from Aristotelian thought. According to this schema, plants lacked the capacity for intentional motion, could reproduce through cuttings, and passively derived their nutriments from the environment.[41] Trembley's course of experimentation tested all of these criteria, which in many ways reinforced the organism's enigmatic character. "I am all the more circumspect because the facts which the little bodies present are new," Trembley wrote to Réaumur on February 16, 1741. "I suspend my judgment. I do not dare give it a name."[42]

The self-generated movement of the polyp's threads and its ability to propel itself initially led Trembley to believe it was of an animal nature, even as its anatomical simplicity made it more akin to a plant. He therefore decided to test its method of reproduction, which prompted one of his bolder experiments: cutting the polyps. Animals would assuredly not survive such a procedure and, in fact, Trembley assumed the polyps would die. Yet the procedure only confounded him further.[43] After slicing one of the polyps in half, he discovered that each section regenerated and formed a perfect, complete organism. Equally confounding was the polyp's ability to bud, another method of producing offspring asexually. He observed an "excrescence" on one of the polyps as it grew, developed arms, and finally detached itself as a complete, independent creature. Réaumur proposed that it might be the maturation of an egg attached to the original polyp, but Trembley averred that the offspring shared a membrane with its parent, emerging from the depths of its skin.[44]

Trembley quickly realized that the polyp's skin was its most salient feature, since it seemed to consist of nothing *but* skin. In his reports to the Royal Society, he observed that its body was little more than a single pouch with an orifice at one end, which he discovered served as both its mouth and its anus. In 1742 Trembley described the creature's means of nutrition in a report to the Royal Society:

> Their Arms extended into the Water, are so many Snares which they set for Numbers of small Insects that are swimming there. As soon as any of them touches one of the Arms, it is caught. . . .
>
> A *Polypus* can master a Worm twice or thrice as long as himself. He seizes it, he draws it to his Mouth, and what is more, swallows it whole.
>
> If the Worm comes endways to the Mouth, he swallows it by that End; if not, he makes it enter double into his Stomach, and the Skin of the *Polypus* gives way. The Size of the Stomach extends itself so as to take in a much larger Bulk than that of the *Polypus* itself, before it swallowed that Worm. The Worm is forced to make several Windings and Folds in the Stomach, but does not keep there long alive; the *Polypus* sucks it, and after having drawn from it what serves for his Nourishment, he voids the Remainder by his Mouth, and these are his Excrements.[45]

His final and most famous experiment on the polyp related to the issue of nutrition. Borrowing from a theory proposed by the distinguished Dutch physician

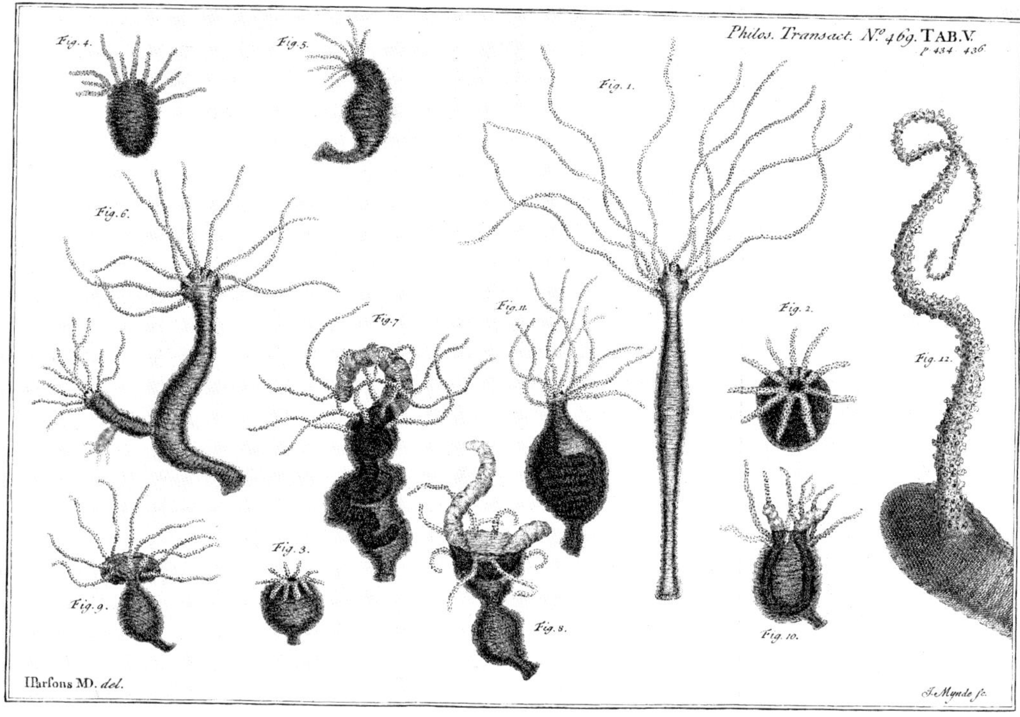

FIG. 3.5. James Parsons (James Mynde, engraver), *Tab. V.* From *Philosophical Transactions* no. 469 (1742/1743). © Royal Society, London.

and botanist Hermann Boerhaave, that "plants have external roots and animals, interior ones," Trembley wondered whether turning the polyp inside out would allow it to absorb sustenance from its environment, like a plant. Using a boar bristle he carefully inverted a polyp, so that "their Inside is become their Outside, and their Outside their Inside." Yet this changed nothing—he noted that "they eat, they grow, and they multiply, as if they had never been turned."[46] For Trembley the polyp's ability to propel itself, its method of asexual reproduction, and its predatory habits all confirmed its animal nature. At the same time, it was an animal like none he had ever witnessed, without viscera, muscles, or even a central nervous system.

If the polyp challenged long-held criteria for classifying life, its capacity to contract and extend, sprout offspring from its body, and even invert presented challenges for its visual representation. A protean creature, it could shift shapes

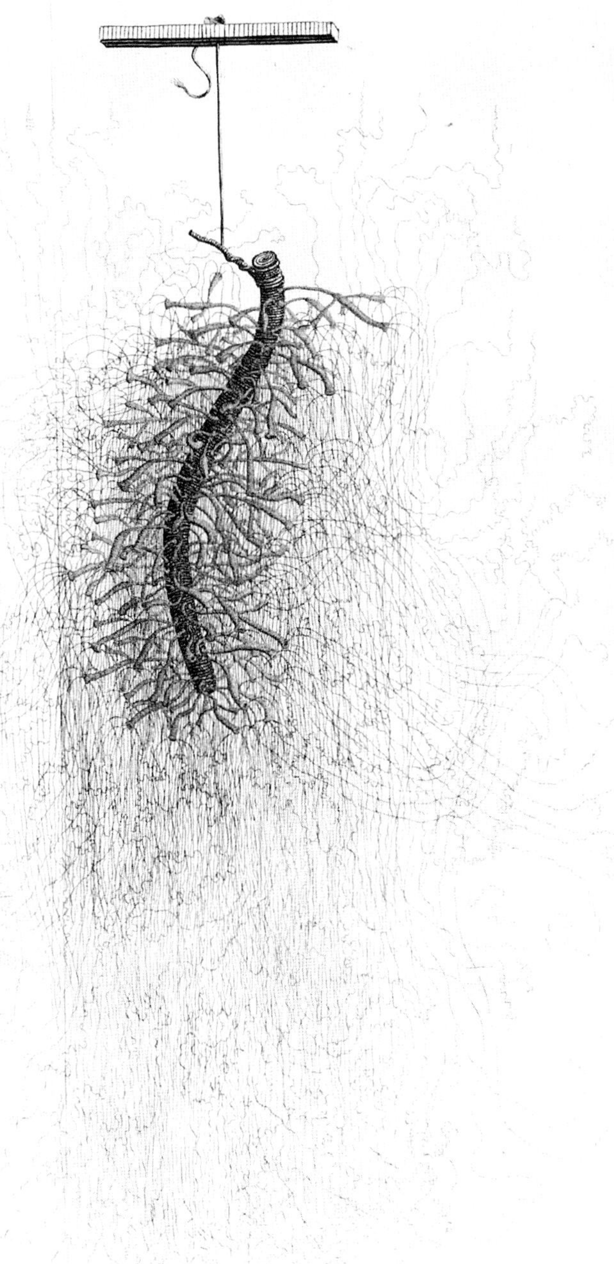

Fig. 3.6. Pierre Lyonnet,
Plate 9, Mem. 3. From
Abraham Trembley,
*Mémoires pour servir
à l'histoire d'un genre
de polypes d'eau douce*
(Leiden, 1744). © Royal
Society, London.

through contraction, extension, and the growth of new appendages and even upset the distinction between inside and outside. Visual depictions of the polyp by necessity rejected the principles of transparency and self-evidence so common to natural history illustration. Rather than combine numerous views into a sort of visual average, representations of the polyp often comprise sequences of observations, in an attempt to capture the polyp's mutability. A plate published in the same issue of the *Philosophical Transactions* as Trembley's report features numerous figures that depict the organism extended and contracted, sprouting excrescences, and swallowing worms twice its size (fig. 3.5).

Trembley's *Mémoires pour servir à l'histoire d'un genre de polypes d'eau douce* (1744) likewise included sequenced figures to illustrate how the polyp moved and changed form, yet the book's most compelling illustration somewhat alters this format (fig. 3.6). Although it shows a multiplicity of figures, they are not in sequence. Instead, the plate depicts a small branch suspended by a string and covered in clusters of long-armed polyps, whose bodies are rendered with the same heavy lines as the wood.[47] Aside from the slightly bell-shaped form of the polyps, they are almost indistinguishable from the string-wrapped twig that protrudes from the top of the branch. The wood and polyp bodies are surrounded by a vibrating halo of thin jittery lines that represent the polyps' arms. These lines do not just drape downward but reach both out and up, indicating their intentional movement. This linear cacophony eschews the various rationalizing devices of natural history illustration. Instead, it presents an unexpectedly mutable natural world, one in which a small branch of wood appears vigorously alive, spilling with so many sensible appendages.

Trembley's research on the polyp was exceptionally well known because of his reports in the *Philosophical Transactions*, his *Mémoires*, and the publications of others. Beyond merely describing his experiments, his reports invited readers to re-create them. "If any Persons in *England* shall be desirous to make Observations on the *Polypus*, and to repeat my Experiments; I hope I shall be able to send some over, in case they shall not be found there," Trembley declared. "I am ready to impart to every one who shall desire to make Observations on these Animals, all the Means and Contrivances I have used."[48] Mark Ratcliff refers to this as Trembley's "strategy of generosity," an approach to scientific research that in some ways parallels the theories it encouraged by replacing a top-down, hierarchical model with a dispersed and dynamic network. In the mid-eighteenth century the organized body and the study of that body were not necessarily centrally controlled but, rather, were the accretion of disparate contributors.[49]

Henry Baker was one of those naturalists quick to investigate the polyp, even rushing to get his own book, *An Attempt Towards a Natural History of the Polype* (1743), into print before Trembley's *Mémoires* could be published.[50] He offered up the Royal Society's motto—*nullius in verba*, take no one's word—as his explanation. "In Cases of an extraordinary Nature, the Experiments and Attestations of different People serve more effectually to establish Truth," he observed, and his own re-staging of Trembley's polyp studies allowed him to affirm their accuracy. As Baker assured his readers, "real Facts are incontestable Arguments" and hold their own against any ideas or theories regarding the order of nature, regardless of how entrenched they might be.[51]

As naturalists such as Baker were re-creating Trembley's studies of the polyp, others were developing new ones that interrogated the vitality of organic matter. The microscopist John Turberville Needham, for instance, conducted a series of experiments to test the assertions of Georges-Louis Leclerc, comte de Buffon, that a *"prima Stamina"* permeated all bodies, animal or vegetable. Although Needham originally hewed to Cartesian dualism and a mechanistic worldview, he set out to test this premise.[52] His initial experiment involved boiling mutton gravy (the "pure unmix'd Quintessence . . . of an animal Body"), sealing it in a phial, and observing whether any microscopic animalcula developed over the course of several days.[53] According to Needham's hypothesis, the development of animalcula in this supposedly uncorrupted quintessence would indicate the existence of a vital force responsible for the generation, organization, and movement of matter. He repeated this procedure with plant infusions and reported his findings in a 1748 issue of *Philosophical Transactions*. The phials "swarm'd with Life," he remarked, indicating that "all Substances animal or vegetable, and in every Part of those Substances, as far as the smallest microscopical Point," possessed a living force.[54]

Needham's experiments were flawed—he had neither boiled the infusions long enough to kill existing microbes nor taken efforts to protect them from the open air before sealing the phials—yet such "proof" of matter's inherent vitality led some to believe that there existed no clear distinction between plant and animal life. In another issue of the *Philosophical Transactions*, a gentleman from Cambridge observed, "Now in such a Scale, who is the Man that will be bold to say, Just here Animal Life intirely ends, and here Vegetable Life begins?

Or, Just thus far, and no further . . . ? Or, again, Who will venture to say, Life in every Animal is a Thing absolutely different from that which we dignify by the same Name in every Vegetable?"[55]

Bartram was patently aware of these debates regarding mind and matter, which thread through the introduction to his *Travels*. Bartram began the book with a mechanistic description of plant life that reads as relatively doctrinaire, yet he quickly adopted a vitalist tone. He detailed, for instance, the morphology of the *Sarracenia* or pitcher plant, which consists primarily of funnel-shaped leaves that collect rainwater for sustenance. The water is absorbed into the plant through imperceptible pores and is then conducted through it by a complex system of vessels and ducts, allowing the *Sarracenia* to withstand periods of heat and drought. He noted that pitcher plants are insect catchers; their water-filled leaves lure unsuspecting flies and bees, which become trapped by the stiff hairs in the leaves' interior. The plant, he seemed to suggest, functions simply by physics, rather than by any inherent natural perception.[56]

In the next paragraph, however, he complicated this mechanistic view with a collection of exclamations: "But admirable are the properties of the admirable *Dionaea muscipula*! . . . Astonishing production! see the incarnate lobes expanding, how gay and ludicrous they appear! ready on the spring to intrap incautious deluded insects, what artifice!"[57] In this passage he described the newly discovered Venus flytrap, which was brought to the attention of European virtuosi in 1759.[58] During a period when naturalists were probing the divisions between animal and vegetable, the flytrap's ability to pursue insects actively (rather than catch them passively, like the *Sarracenia*) made it an especially confounding organism, not unlike the polyp. Peter Collinson encouraged the Bartrams to send him the plant, which they had successfully cultivated at their garden by August 1762. Unfortunately, their attempts to send living specimens abroad were in vain; they could only provide seeds and a dried leaf and flower, which Collinson received later that fall. The living flytrap was not accessible in Europe until the summer of 1768, when the Pennsylvania botanist William Young successfully transported more than a hundred plants from the colonies.[59]

Working with a live specimen, the naturalist John Ellis meticulously described the plant according to the Linnaean system, confidently declaring the Venus flytrap an entirely new genus and species. In a letter to John Bartram, Collinson indicated that mere textual description had been inadequate for conveying the plant's singularity, explaining that "it was impossible to Comprehend

It from any Description made Mee."[60] Yet it was equally impossible to comprehend the plant fully from dried specimens, which altered the plant's shape and foreclosed the possibility of testing its sensitivity. It was only after the arrival of Young's shipment that Ellis was able to publish his botanical description, initially as a report in a September 1768 issue of the *St. James Chronicle* and later as an appendix to his *Directions for Bringing Over Seeds and Plants, from the East-Indies* (1770). Included in the book was the first published illustration of the Venus flytrap, though not its first known visual depiction. Bartram had already portrayed it in his drawing of the American lotus but, since no one in Europe had seen a living specimen, its inclusion was described as fancy.[61]

As part of his enduring interest in plantlike animals—known as zoophytes—and animal-like plants, Ellis endeavored to clarify the status of the Venus flytrap. He had already conducted microscopic studies of corals, by which he affirmed their animal nature, and published his findings in *An Essay Toward a Natural History of the Corallines* (1755). He would expand his research to include other forms of plantlike animals, such as sea anemones and sponges, which was posthumously published in his *Natural History of Many Curious and Uncommon Zoophytes* (1786). Notably, his lifelong interest in these organisms stemmed not from a fascination with the possible porosity of taxonomic categories but from a desire to establish them and affirm their validity.[62]

After examining the living flytrap, Ellis seemed initially to have entertained some doubt regarding its classification. He was uncertain whether its predatory movements indicated a sort of perception, especially if it drew nutriment from the insects it caught. In October 1768, he wrote to the Scottish physician David Skene, "Lord Moreton asked me a shrewd Question, . . . Do you think Sr that the plant receives any nourishment from Insects it catches? I acknowledge'd my ignorance."[63] By 1770, however, Ellis's investigations had allayed his concerns. In the description appended to his book on the transportation of plants, Ellis maintained that the flytrap's capture of insects did not reveal any form of sentience since its leaves were unable to "distinguish an animal, from a vegetable or mineral, substance; for if we introduce a straw or a pin between the lobes, it will grasp it full as fast as if it was an insect."[64] Adopting a counterintuitive stance, Ellis held that the Venus flytrap's predatory movements actually affirmed its lack of sentience and its vegetable nature. Other naturalists were more circumspect. Even Linnaeus, the arch systematist of the eighteenth century, suggested that some organisms, such as fungi, transmuted between vegetable and animal over the course of their life cycle. He did not place the flytrap in this category but, rather, referred to it as a "miracle of nature" for its combination of plant and animal qualities.[65]

In *Travels*, Bartram indicated a propensity toward vital materialism in his description of the Venus flytrap. "Can we after viewing this object, hesitate a moment to confess, that vegetable beings are endued with some sensible faculties or attributes, similar to those that dignify animal nature[?]" he asked. "They are organical, living and self-moving bodies, for we see here, in this plant, motion and volition." To underscore the arbitrariness of the distinctions among kingdoms, Bartram proceeded with an assessment of other plants that exhibited the same kind of "motion and volition," like the creeping vines of genera *Cucurbita*, *Momordica*, and *Vitis*, or gourds and grapes. He observed how the tendrils of these plants seem possessed of a combination of sight and touch and wondered whether it is sense or instinct that guides them.[66] His question harks back to Haller's separation of sensibility and irritability, but for Bartram this was ultimately a false distinction. Like the plant physiologist Thomas Percival, who maintained that "the idea of irritation involves in it that of feeling," Bartram seems to have held the view that the two were entwined.[67] In Bartram's view, sensibility and irritability, sense and instinct were, in their most basic forms, indistinguishable.

One of Bartram's first drawings for Fothergill, of the *Momordica charantia* or balsam pear (fig. 3.7), illuminates his belief in vegetable sensibility. It depicts three of the plant's fruit at different points of maturation—ranging from a tiny tight pear near the top of the drawing to the bursting one below—surrounded by spooling tendrils and waving leaves. Amy Meyers observes how the curving and twisting lines in Bartram's drawings lend his depicted objects a sense of animation, and this composition goes even further. The many coils and spirals suggest that the plant is self-moving, rather than animated by an exterior force.[68] As Alexander Nemerov notes regarding the self-aware quality of the fruit and vegetables in Raphaelle Peale's still lives, Bartram's *Momordica* is consciously oriented toward the viewer, as though willfully placing itself on display.[69] The large pear, with its broken, peeling skin, seems almost to disrobe, lifting its pericarp to reveal the plant's seeds. The pitching leaves appear to flag the viewer's attention.

Bartram's description on the back of the drawing underscores the balsam pear's uncanny sense of agency. His text initially describes the plant as an inert

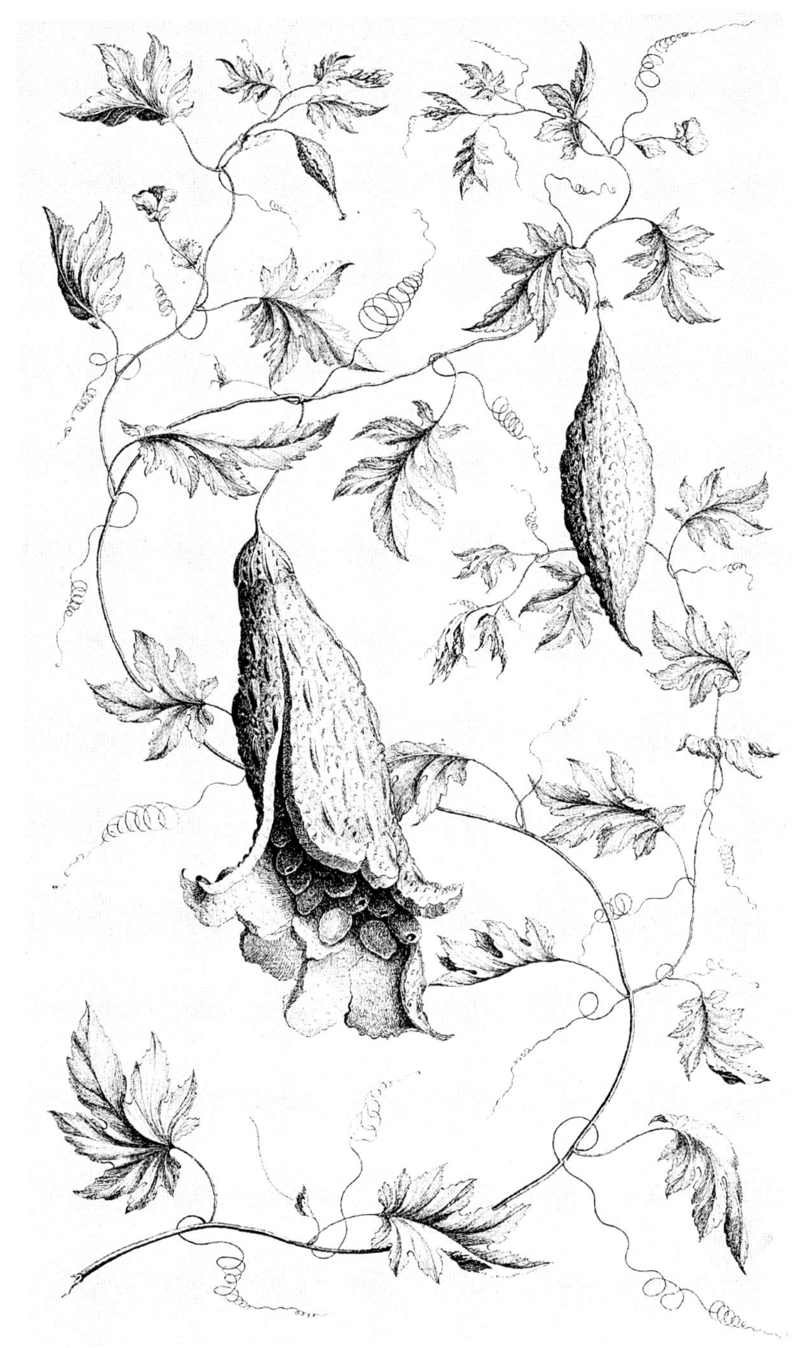

Fig. 3.7. William Bartram, *Species of Momordica* [*Momordica charantia* or balsam pear], ca. 1769. © Natural History Museum, London.

object, collected and distributed: Bartram received "a branch with fruite to make a drawing" from the garden of Thomas Penn, then the colonial proprietor of Pennsylvania, who had obtained its seed from the banks of the Illinois River. The *Momordica*, however, soon becomes a subject that acts, as opposed to an object acted upon. The passage describes how its fruit, "when ripe & ready," splits open "with great elasticity[,] disclosing a scene of still greater beauty & wonder."[70] The active quality of the plant—no longer disclosed but disclosing—transforms the relationship between observer and observed as much as it blurs divisions between vegetable and animal life. The *Momordica's* beauty and wonder is accessed more readily through its gesture of self-presentation than through the naturalist's probing interference.

The Problem of Knowing

By challenging rigid distinctions between mind and matter, eighteenth-century microscopic studies also influenced conceptions of the naturalist and of what the naturalist could reasonably know. Despite the new worlds opened up to view through the magnifying lens, the naturalist's understanding was in some sense obscured rather than clarified. Baker, in his book on the polyp, observed that just because some things may be plain to the senses does not necessarily mean we wholly comprehend them. To assume otherwise only leads us into "an unfathomable Abyss . . . and [we] are in the utmost Danger of Shipwrecking our Understanding."[71] It was a caution that Bartram embraced. If objects accessible via the senses were beyond our understanding, then the greater part of the world could only be more so. Nature, its origins and order, were "invisible, incomprehensible, how can it be otherwise? when we cannot see the end or origin of a nerve or vein, while the divisibility of mater [sic] or fluid, is infinite."[72] Bartram firmly believed in the existence of a divine natural order, but he also believed in the futility of trying to discern it in all its complexity. As a result, his natural history shifts emphasis from the categorization of species—which could always only be conjecture—to an examination of their lived relationships and metamorphic processes.

Fothergill noted receipt of Bartram's balsam pear in a letter from January 1770; his rendering was included in a shipment from Philadelphia that contained two casks of botanicals, two snakes, a pair of bullfrogs, and five additional drawings.[73] Not all of the five additional works are still extant, but the two that remain disrupt the taxonomic thrust of eighteenth-century Anglo-American natural history, much like Bartram's drawing of the *Momordica*. They include a watercolor rendering of a seed vessel of the American lotus, tucked into a tableau with a snake consuming a writhing frog (plate 4), and another featuring a spotted turtle (*Clemmys guttata*) and two entwined snakes (plate 5). Although these compositions were intended as specimen drawings, they are quite unlike the Linnaean models Fothergill would have anticipated. Their purported subjects seem secondary to the drawings' emphases on temporal change and lived relationships, of which the naturalist is inherently a part.

For instance, the drawing of the lotus seedpod portrays a cluster of organisms with different life spans, and it depicts them at various points in those spans. The blackroot on the left displays its pale inflorescence, which lasts several months, while the dragonfly that hovers above lives but eight weeks. On the right, a ruby-throated hummingbird—living up to nine years—perches above the dried lotus fruit. The fruit begins as a small, tender cone at the center of the lotus flower but, over the course of six weeks, doubles in size. The flower's petals fall away, and the fruit's seeds mature over the course of another six weeks. Bartram depicted it at the culmination of its three-month development, as its desiccated husk finally spills its seeds on the earth. The seed vessel indicates emergent life, just as the frog and snake portend death, echoing the vibrant life in decay of the magnified view.

The conjunction of different temporal durations in the drawing recall the disturbing relativity introduced by the microscope. As Baker indicated in *The Microscope Made Easy*, the concept of duration—like concepts of size and space—is comparative, deriving from each organism's own experience.[74] Implicit here is the idea that plants and animals have their own experiences, which compete with those of naturalists, whose perception of duration will never align with the specimens they observe. Indeed, Baker maintained, the naturalist could never gain a sense of the perceptions, judgments, or understanding of another organism: "May not other Creatures, whose Structure, Organs and Way of living bear no Resemblance to ours, . . . have Sensations also different from ours, not only in Degree but Kind, for their particular Security and Happiness?"[75]

Like the drawing of the lotus fruit, Bartram's rendering of the spotted turtle and snakes points to the vexed position of the naturalist, while also conjoining prospective life with imminent death. It depicts a female turtle—identified by her long thin tail and yellow neck markings—trundling away from the edge of a pond, possibly to lay her eggs. Her actions contrast with the portrayal of the entwined snakes just below, where one snake consumes another. The drawing's construction of space heightens this unsettling juxtaposition: the snakes are located on a small hummock and are viewed from the side, while the turtle and pond are depicted from above. As with Bartram's *Great Alachua-Savana, in East Florida*, the drawing conjoins an aerial perspective—suggestive of an all-encompassing prospect—with a ground-level view. The observer inhabits a liminal position, neither fully outside nor inside the scene. As this drawing makes clear, the desire for a comprehensive overview of the cycles of life and death is confounded by the inability to extract oneself from those cycles. The letter that accompanied the drawing, written by John Bartram, reaffirms this point. "I have sent thee . . . a couple of small snakes which my son William found just before I began to pack," he explained to Fothergill. "[T]he biger one was swallowing the lesser as thee may see in his draught he pulled it out of the snakes mouth with some dificulty for fear of hurting either."[76] As John's letter attests, William was as much a participant in the scene as he was an observer.

The depiction of the two snakes points to another theme that tracks through the drawings: the act of consumption and digestion. As a metamorphic process that transformed natural diversity into unity, digestion was a key eighteenth-century trope for describing the synthesis of facts into knowledge. Yet, as Daniel Cottom points out, this process was also troubling. On one hand, it stood as an eminently rational process, in which subjective sensory data fused into a totalizing schema, but on the other, it seemed utterly fantastical, creating uncategorizable fusions out of distinct objects. Knowledge depended on digestion and metamorphosis, yet it was also threatened by that process. The totalizing narrative that emerged from digestion was always in danger of devolving into chaos.[77] Bartram's drawings for Fothergill speak to this danger.

In the spotted turtle drawing, the entwined snakes suggest a revision of the ouroboros, the snake that devours itself. Although the ouroboros would come to symbolize the unity of the thirteen colonies against Britain, here its meaning is arguably less political.[78] The fact of one snake ingesting the other, an act that reads almost as cannibalistic consumption, challenges the distinctions between sameness and difference or, at least, reveals the arbitrariness of these distinctions.[79] According to John Bartram, after William pulled the snakes apart "thay both crawled about in a bucket without any hostility for many hours."[80] That one snake should attempt to swallow the other and, later, peacefully coexist upsets any notion of stability; there could be no telling what Fothergill would find when opening his shipment of natural productions. And, indeed, he did not tell. Although his January 1770 letter registers his receipt of the botanicals, bullfrogs, and drawings of the balsam pear and lotus fruit, he makes no mention of the snakes or their portrayal.

The void in his letter seems peculiarly apt since the cannibalistic figure was ominously described as a "gulf and an abyss" in Pierre Bayle's popular *Dictionnaire historique et critique* (1696; first English translation, 1709). In Cottom's analysis, the cannibalistic figure in the early modern imaginary initiated a metamorphic process with "no beginning or end in nature, culture, or spirit," serving as an emblem of an incoherent and unstable world that consumes itself.[81] But what makes Bayle's assessment of the cannibal especially compelling is not just its summation of the potential threat of indiscriminate consumption and metamorphosis but its position in the *Dictionnaire*. The book's essays are larded with extensive "remarks" or footnotes, and this is precisely where his assessment of the cannibalistic figure is located, quite aptly in the entry on Ovid. The main text offers an overview of Ovid's life, from his recognition as a poet to his exile from Rome, but the footnotes sag under the weight of the many references and the reflections they contain. The discussion of cannibalism occurs in footnote G, which runs more than six folio pages. In this excessively long aside, Bayle asserted that the chaos in Ovid's *Metamorphoses* never truly resolves: "Living bodies subsist only on destruction: everything which serves for the support of their life, loses its form, and changes its state and species."[82] Yet at the same time he attempted to illuminate the harrowing world of endless transmutations in Ovid, he also cordoned it off from the broader narrative, relegating it to an "undigested" supplement.[83] By isolating such details in the footnotes, Bayle created a bulwark against potential disorder, even as he described it.

The monitoring of digestion and metamorphosis in Bayle's *Dictionnaire* takes visual form in Mark Catesby's *Natural History of Carolina, Florida, and the Bahama Islands* (1731–1743). His etching of rice birds (visually quoted in William Bartram's drawing of the marsh hawk and in Edwards's etching after it) depicts a hen on a bowed stalk of rice, delicately holding a grain in her beak,

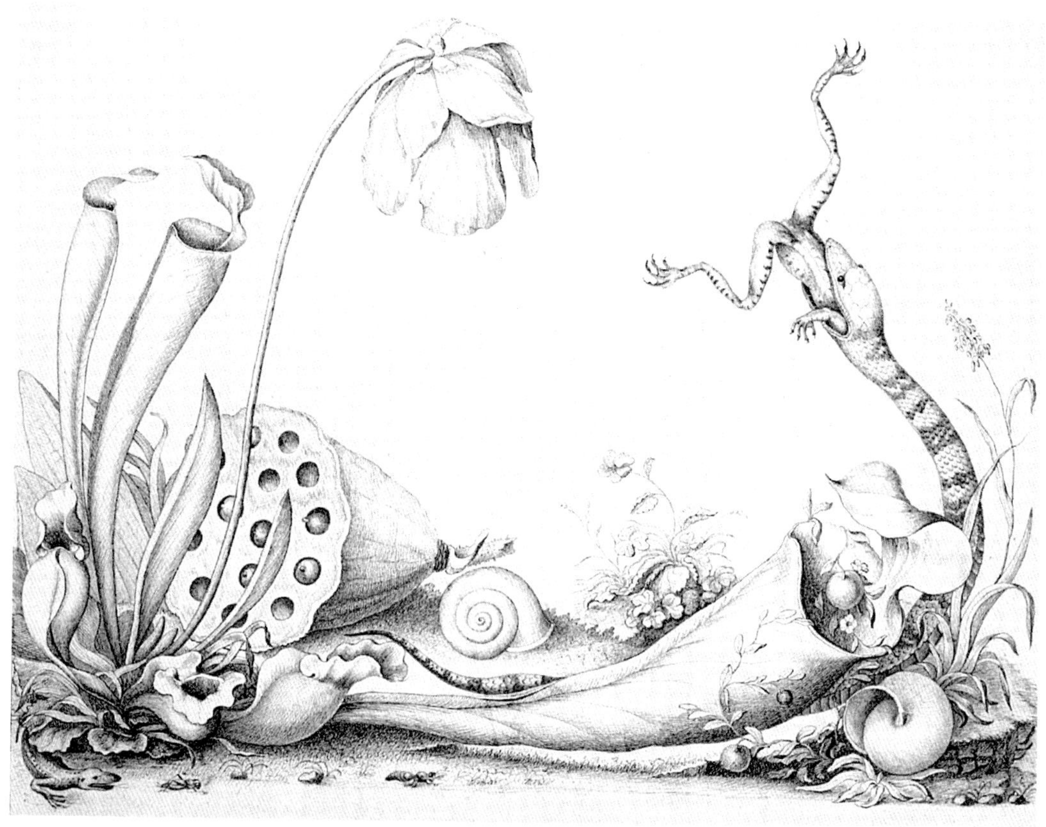

FIG. 3.8. William Bartram, *Sarasena and Seed Vessell of Calocasia* [*Sarracenia* or pitcher plant, and *Nelumbo lutea* or American lotus], ca. 1767. © Natural History Museum, London.

while the cock—perched below on a small half circle of earth—gleans rice from the grass (see fig. 1.4). Birds and plant join together in a firm pyramidal composition that suggests a stable and accessible natural order. The birds engage in an act of consumption that appears visually controlled and coherent, rather than the confusingly entwined snakes of Bartram's drawing. Although Catesby in the accompanying text pointed to the threat of unbounded consumption in his reference to the birds' gluttony, he also attempted to moderate this disruption. Although the birds' gorging allows them to clear a forty-acre crop, it also rendered them "so excessive fat, that they fly sluggishly and with difficulty" and are easy prey.[84] By decimating the planter's rice crop, he explained, the birds become delicious fat fowl for him to consume. The potential unsettling of a stable natural order is set right by the birds' transformation into a delicacy.[85] Although Catesby recognized the disruption to local ecologies that was caused by transplantation, his verbal and visual descriptions of the rice birds seem intended to reinforce a sense of nature's stability and order.

Bartram's drawing of the lotus seedpod rejects the visual stability of Catesby's rice birds. It takes as its model a drawing likely made for Collinson, which similarly features a lotus fruit and a snake consuming a frog (fig. 3.8). In the Collinson version, the composition is flipped left to right, and two species of carnivorous *Sarracenia* replace the blackroot. Meyers describes this work as one of Bartram's most compelling portrayals of an entwined organic world. The voracity of the snake and gap-mouthed plant are balanced by the indication of fecundity and plenitude of the cornucopia-like *Sarracenia* leaf in the drawing's foreground. The drawing lacks the pyramidal stability of Catesby's rice birds, yet it still presents nature as tidily organized—if dynamic and potentially hostile—through the contained oval composition and the formal mirroring of snake and plant.[86]

Bartram's drawing for Fothergill, however, is far more compositionally dispersed. In breaking the close circularity of the earlier version for Collinson, it portrays a natural world that is far more uncertain and no less hostile (plate 4). This later version seems almost to parody the order found in Catesby. The stable pyramidal composition of Catesby's illustration is upended in Bartram's conelike fruit, which lolls about on its side in the drawing's foreground. The work offers no clear visual anchor but instead presents a group of plants and animals whose lived relationships are ever in flux. Should the frog survive, would it eat the dragonfly? Might the snake soon pursue the hummingbird? Will the hummingbird feed on the blackroot's inflorescence? Such unpredictability creates an image of the natural world that seems infinitely complex, even cruelly arbitrary.[87]

Through the later drawing's formal dispersion and conceptual instability, Bartram shifted the viewer's attention from the specimens' taxonomic characteristics to the contingent and fluid interactions among them. This move lays bare the vexing search for definitive order and meaning, since securing such order demands a firm conceptual ground on which to stake that claim.[88] In Catesby's description of rice birds, that ground is constructed visually through the pyramidal composition and narratively through the figure of the Carolina planter; it is ultimately a representation of a hierarchical order, in which rice feeds fowl and fowl feeds human. Catesby's etching and letterpress carefully patrols the processes of digestion and metamorphosis in their creation of a great chain of consumption, whose links are all carefully demarcated—rice, bird, human—a chain that Bartram's drawing refuses. In his drawing of the lotus fruit for Fothergill, Bartram portrayed each organism as terrifyingly available for consumption.

The drawing's disorder is further reinforced in the figure of the startlingly hybrid snake-frog. Neither two specimens nor one, the conjoining of snake mouth and frog legs depicts two species in the process of *becoming* one. Hybridity, not surprisingly, undermines the validity of any sense of order that the naturalist might want to impose. This passage in Bartram's drawing recalls the polyp's processes of consumption, as it envelops large worms by folding them in half, leaving their ends to protrude and flail outside the polyp's body. "When an Animal is reported to swallow Creatures larger than its own Size, an Explanation is highly necessary," noted Henry Baker in his *Attempt Towards a Natural History of the Polype*. He attributed this occurrence to the polyp's pliability and capacity for extension, observing that "Something of the like Kind may be observed in Snakes and Lizards, when they swallow large Frogs, Toads, &c."[89] The protean forms of polyps and snakes allowed for the creation of such disturbing fusions.

The snake-frog combination calls to mind Ovid's *Metamorphoses*, a work that details the flux of identity, whose protagonists turn into stags, birds, and spiders. As a student in the Academy of Philadelphia, Bartram would have been familiar with the Roman poet's most famous work.[90] Ovid's *Metamorphoses* was used to instruct students in scanning verse, and the poem's themes of mutability and change were underscored by the curriculum's emphasis on its temporal features of stress and rhythm.[91] Bartram likewise adopted contrapuntal temporalities in his graphic work to draw attention to the natural world's hybridity.

Yet if metamorphosis and hybridity challenged the construction of knowledge, so too did excessively rigid distinctions. In writing to the Reverend William Huddesford, Keeper of the Ashmolean Museum, Fothergill dismissed Bayle's *Dictionnaire* for "throwing some of the most material parts of their history into the form of notes." He continued, "Was I to attempt a thing of this kind I would leave nothing as a note but dates, and references to dates, and even not these, if possible to be avoided."[92] If Bayle's lengthy footnotes or Catesby's illustration of rice birds suggest a patrolling of categories—of what constitutes knowledge and what does not—and the cannibal-like consumption in the spotted turtle drawing signals their dissolution, Bartram's rendering of the lotus fruit pursues a middle course. The snake-frog contributes a sense of conceptual instability to the drawing, but it also works to halt that instability by offering a parable of natural law and human limitations that keeps the visual narrative from spinning into disarray.

Inasmuch as the snake's mouth seems a terrifying abyss as it swallows the frog, the scene is also comic. The upturned frog, its feet splayed and legs akimbo, is slapstick, an inversion of order that is as funny as it is disquieting. This marriage of humor and cruelty at the precise point of transformation echoes the *Metamorphoses*, where the whim and caprice of the natural world are underscored by being presented as a "cosmic comedy of manners"—an equally apt description of Bartram's seedpod drawing.[93] Bartram, in fact, appears to have drawn on the visual iconography of the *Metamorphoses* in his depiction of the frog's hindquarters, which resonates with images of Icarus. Wearing wings crafted by his father, Icarus flies too close to the sun, which melts the wax that binds them, and he subsequently falls to his death.[94] Bernard Picart's engraving of the fall for *The Temple of the Muses; or, the Principal Histories of Fabulous Antiquity* (1733)—perhaps the most popular eighteenth-century illustration of this tale—shows Icarus plummeting toward the sea as his wings fall apart (fig. 3.9). His legs kick clumsily against the sky, much like Bartram's frog.

In both Picart's engraving and Bartram's drawing, these awkward upturned legs reinforce the disorientation caused by a world of perpetual involutions, yet Bartram could have quoted almost any scene from the *Metamorphoses* to achieve this end.[95] The illustration of Cygnus, for instance, is rife with hybrid figures that more suitably convey a fluid cosmos than does Icarus, who underwent no such transformation. Rather than turn into a bird, he simply donned a pair of wings. A potential motivation for Bartram's visual repurposing of Picart's Icarus illustration is the particular tension in the episode between nature and humanity. Bartram captured nature's tragicomic flux, while also indicating that nature does have an order. Even though humans may not be able to discern the full compass of natural law, they should be on guard against subverting it. Icarus's father, for instance, treads prudently in his construction of feather-and-wax

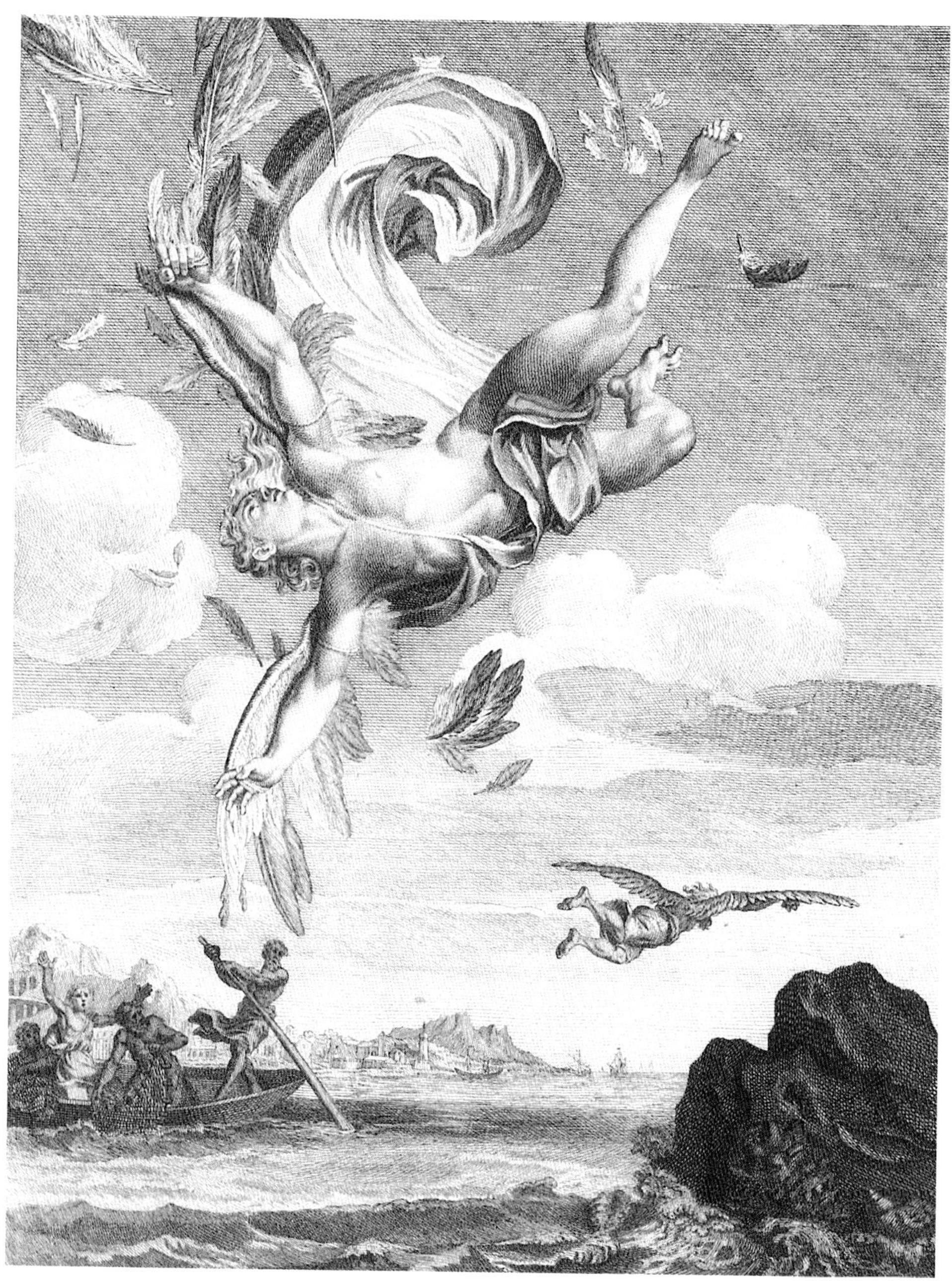

FIG. 3.9. Bernard Picart, *The Fall of Icarus*, detail. From *The Temple of the Muses; or, the Principal Histories of Fabulous Antiquity* (London, 1738). Rijksmuseum, Amsterdam.

wings. He devised a human means for flight, yet he also directed his son "To fly a middle course, lest if you sink / Too low the waves may weight your feathers; if / Too high, the heat may burn them."[96] Icarus failed to follow this advice, a failure that led to his demise. In adapting Icarus's tale for his drawing of the lotus fruit, Bartram countered its seemingly arbitrary acts of consumption by reminding the viewer of nature's order, even though it cannot be fully discerned. More biographically, his reference to Icarus—a son who does not heed his father's advice and suffers grievously—may also have been motivated by his own failures as a planter.

For Bartram, the divine order of nature was immanent in the world around him, even though it fell beyond the scope of human observation. All the aids of optical technology could only reinforce its elusive quality. With this recognition, Bartram concerned himself less with a totalizing taxonomy than with the relativity of perception and the mutability of lived relationships.

In Bartram's *Travels*, Florida's natural springs often fostered not only reflections on representational transparency but also a consideration of the instability of perception. Looking into a vast basin, he exclaimed, "But behold yet something far more admirable, see whole armies [of fish] descending into an abyss, into the mouth of the bubbling fountain, they disappear! are they gone forever?" he asked. "Is it real?"[97] Poised on the threshold of an abyss or another world, Bartram was forced to question his perception as he observed himself observing. The fish transmute in form and size, moving from flies to minnows, as they emerge from the water's depths. They are not wholly separate from the medium in which they swim; just as the fluid bears them aloft, they paint it in a sort of material and aesthetic exchange. The fish scatter and disperse but then reform into kindred tribes, perhaps differently composed than their original armies. Never fully certain whether this variability stemmed from his shifting perceptions or from nature's metamorphic processes, Bartram ultimately refused to distinguish *how* he sees from *what* he sees, refiguring the naturalist as but another element of the natural world.

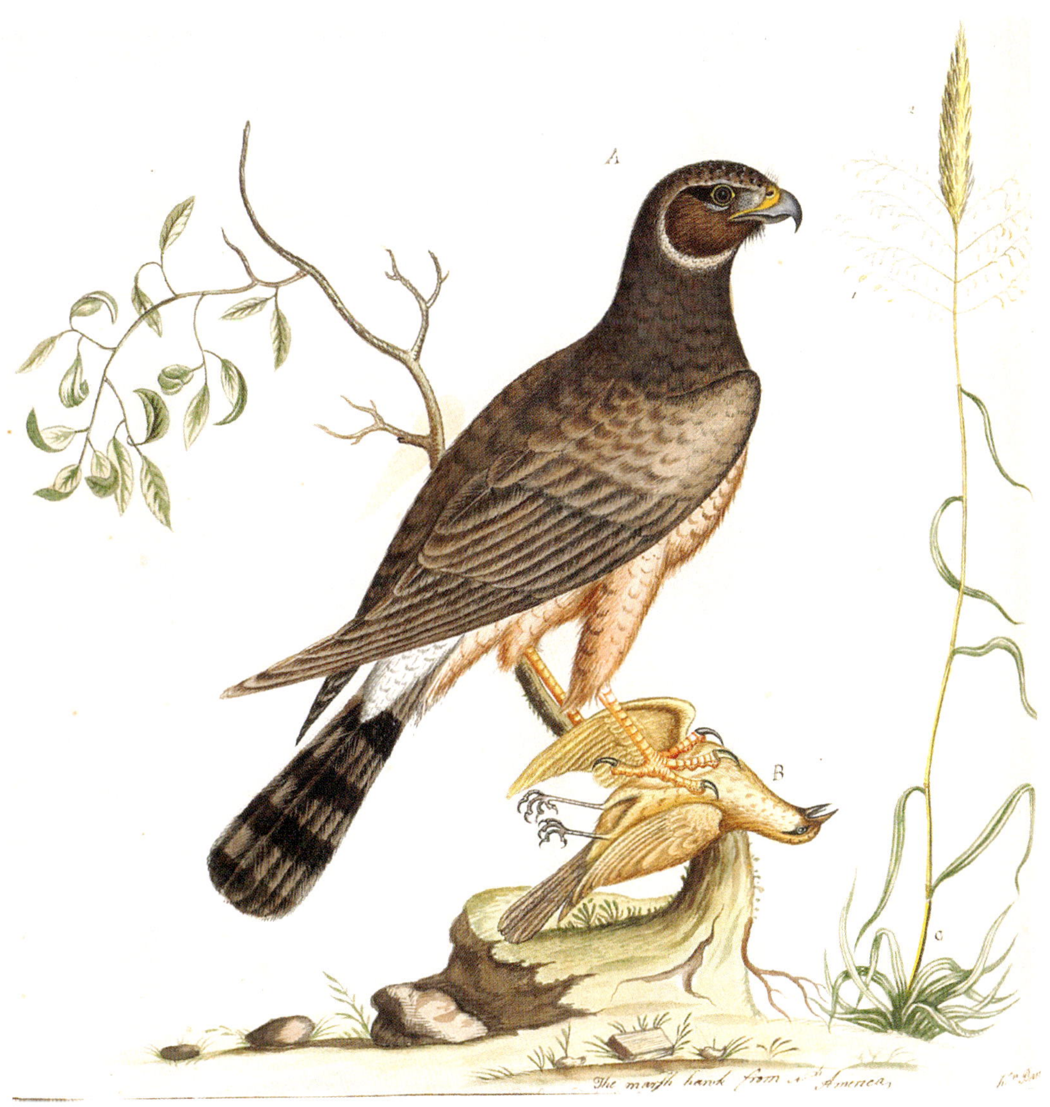

PLATE 1. William Bartram, *The Marsh Hawk from N:th America* [*Circus cyaneus* or northern harrier] *and the Reed Bird* [*Dolichonyx oryzivorus* or bobolink], *the Same with the Rice Bird of Catesby*, 1755–1756. Courtesy of Arader Galleries, New York.

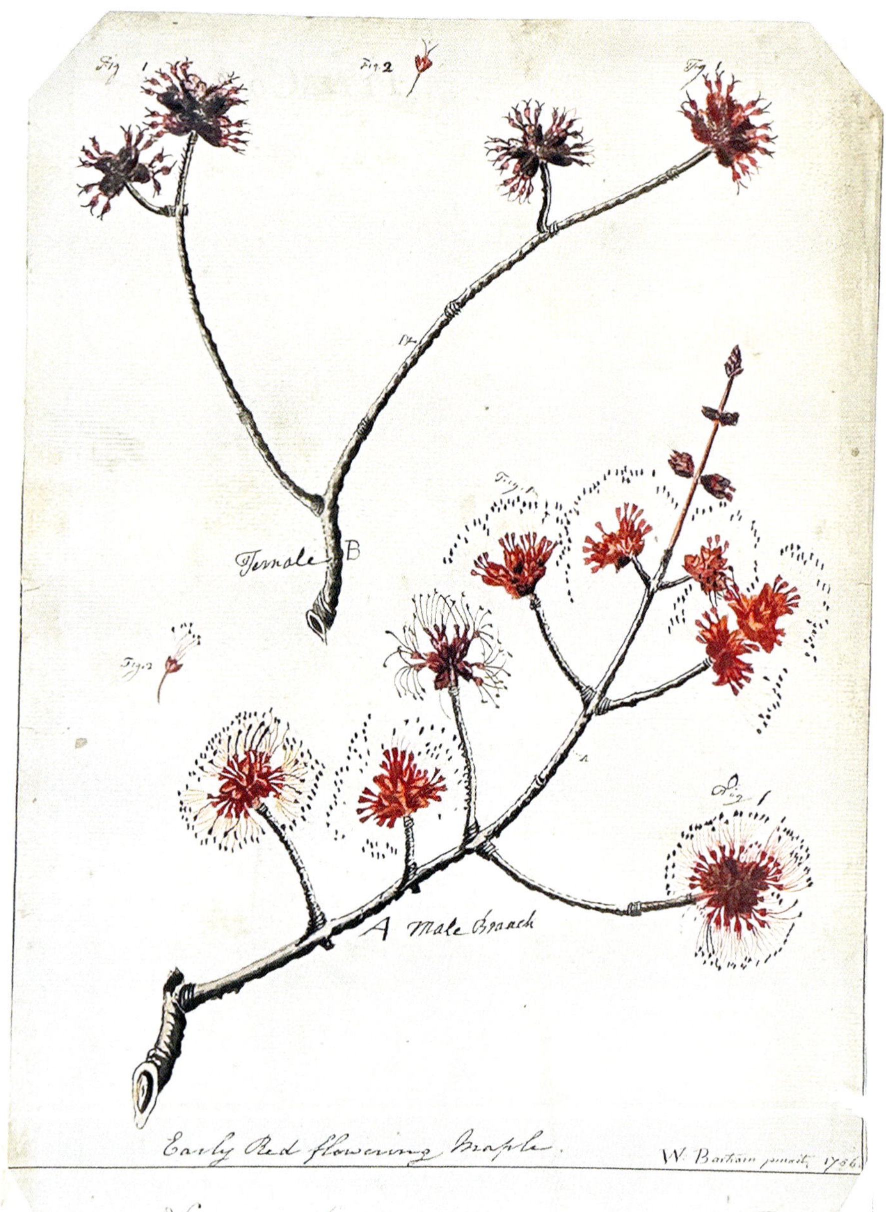

Plate 2. William Bartram, *Early Red Flowering Maple* [*Acer rubrum*], 1756. Courtesy of Arader Galleries, New York.

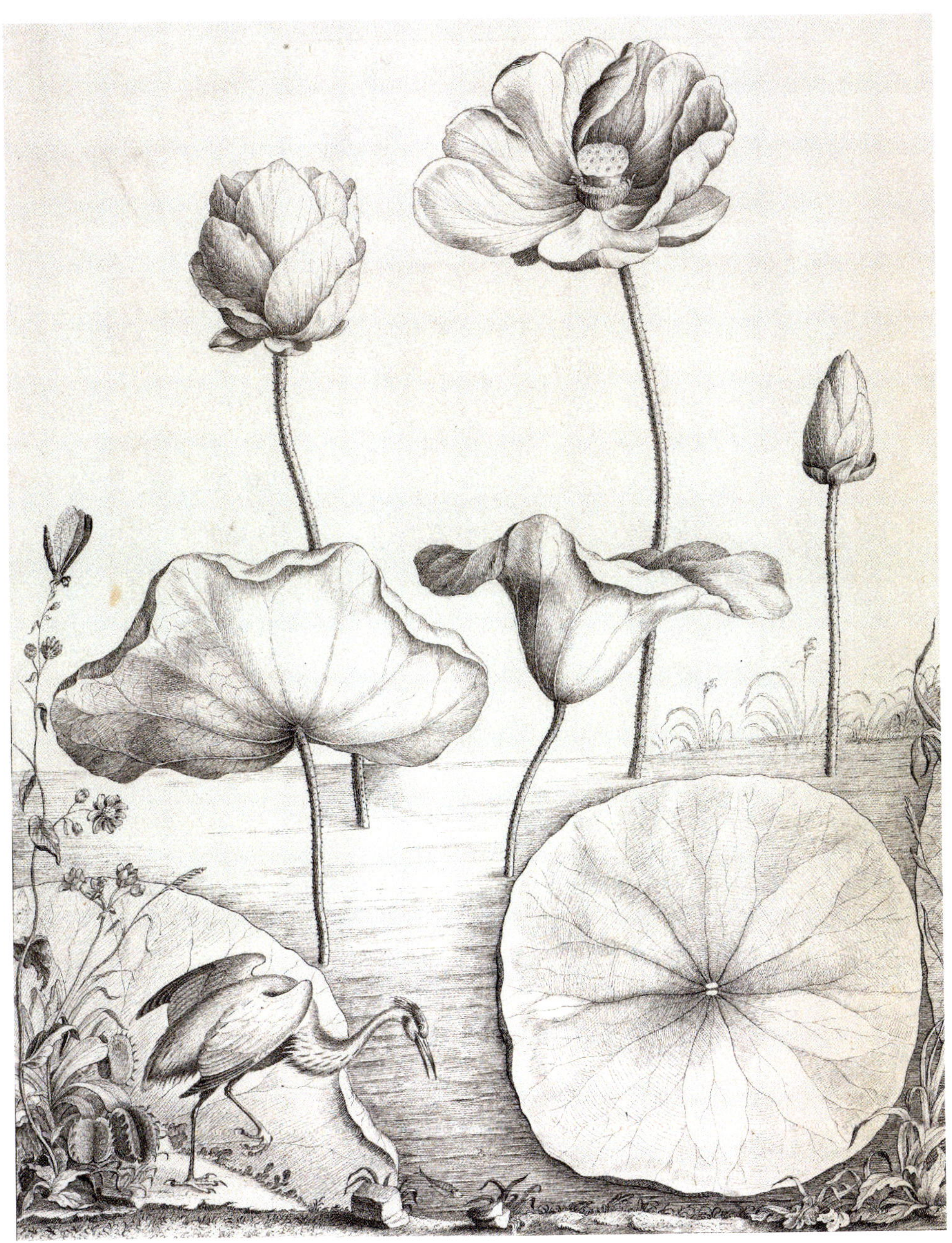

Plate 3. William Bartram, *Round Leafed Nymphea as Flowering*. Colocasia [*Nelumbo lutea* or American lotus], ca. 1767. © Natural History Museum, London.

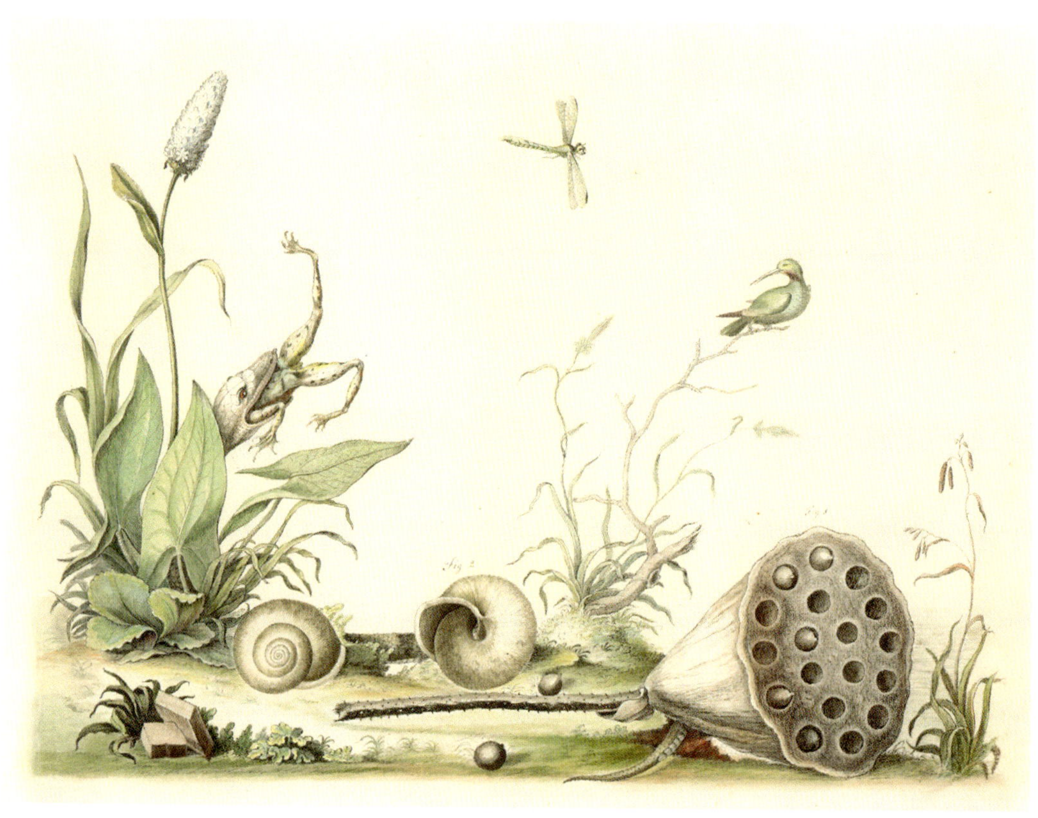

PLATE 4. William Bartram, *Seed Vessell of the Calocasia* [*Nelumbo lutea* or American lotus], ca. 1769. © Natural History Museum, London.

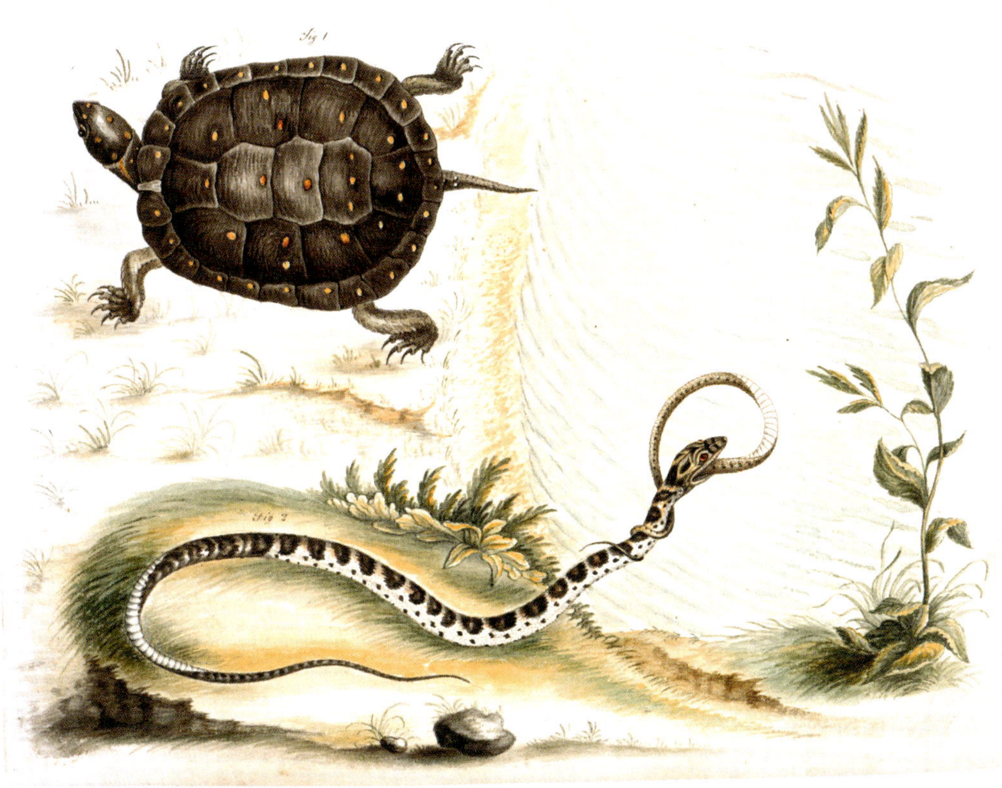

PLATE 5. William Bartram, *A Water Tortoise and Our Wampon Snake* [*Clemmys guttata* or spotted turtle, and *Lampropeltis triangulum* or milk snake], ca. 1769. © Natural History Museum, London.

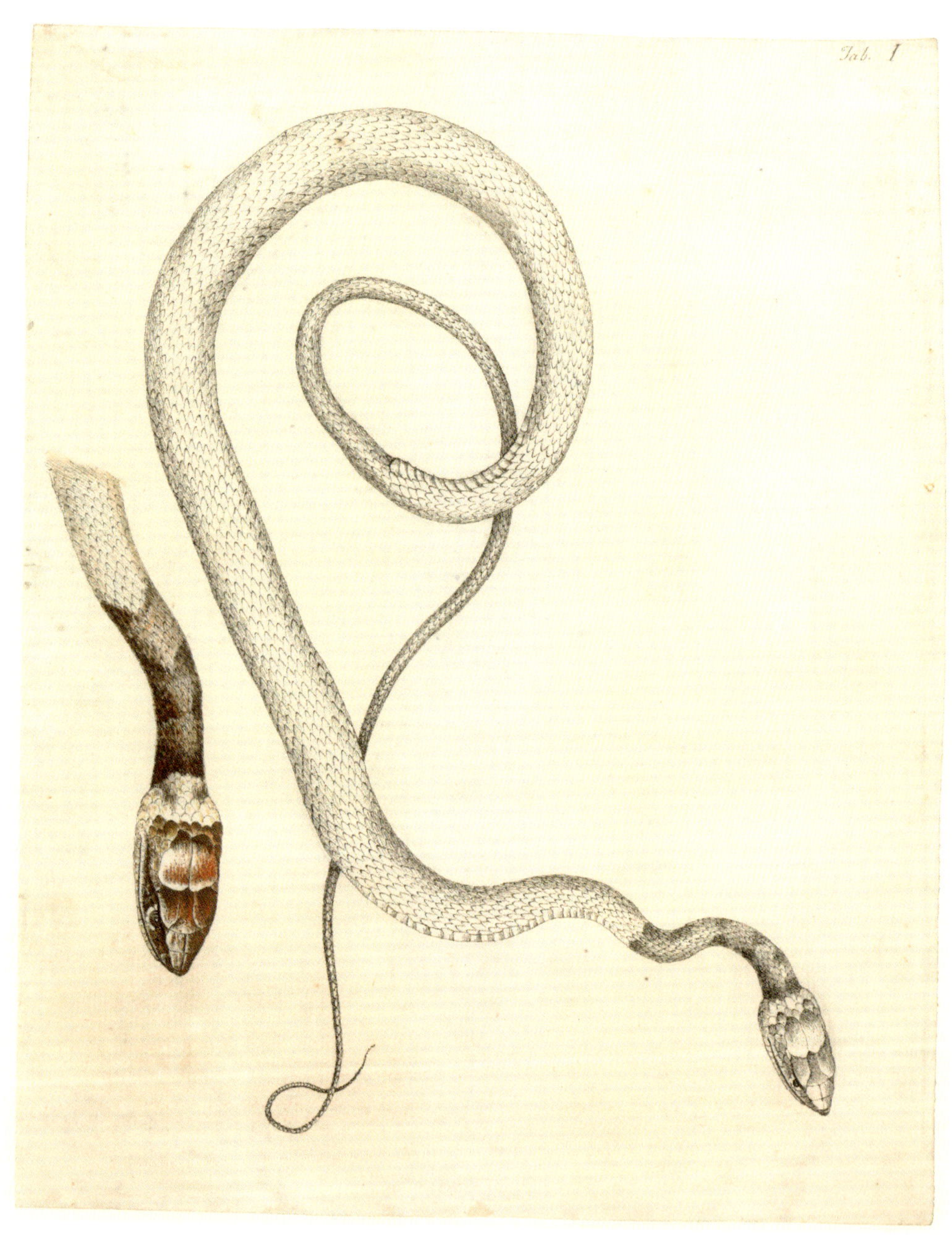

PLATE 6. William Bartram, *Coach Whip Snake [Masticophis flagellum] from Et Florida*, 1774. © Natural History Museum, London.

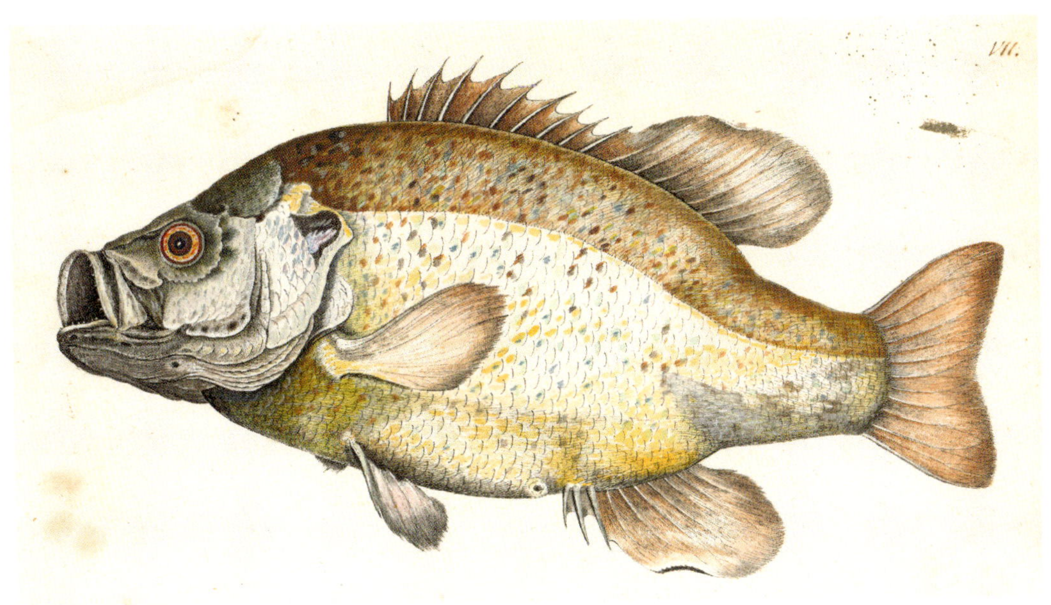

PLATE 7. William Bartram, *Great Yellow Bream* [Chaenobryttus glossus or warmouth] *call'd Old Wife St Johns Et Florida*, 1774. © Natural History Museum, London.

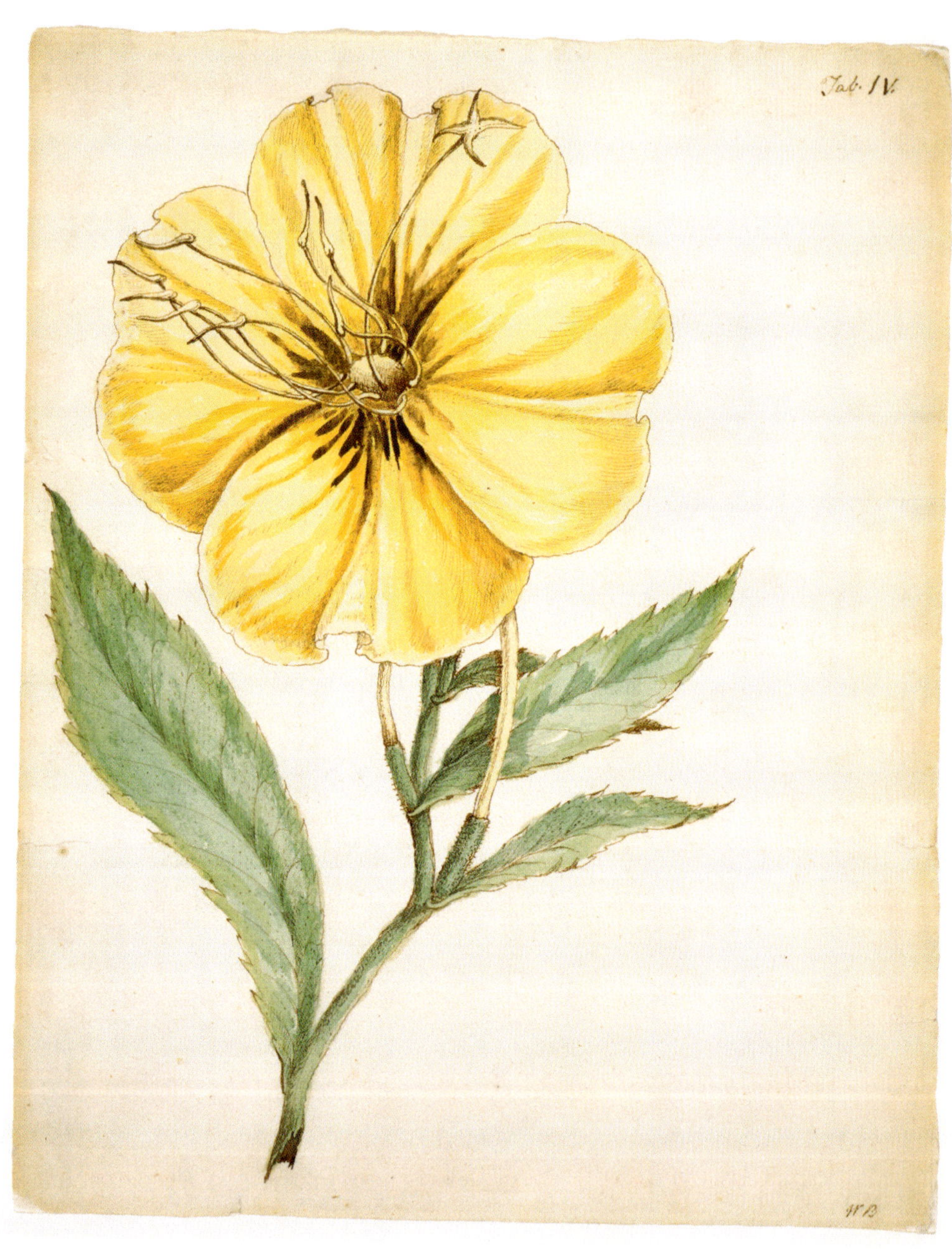

PLATE 8. William Bartram, *Oenothera grandiflora* [Large-flower evening primrose], 1788. © Natural History Museum, London.

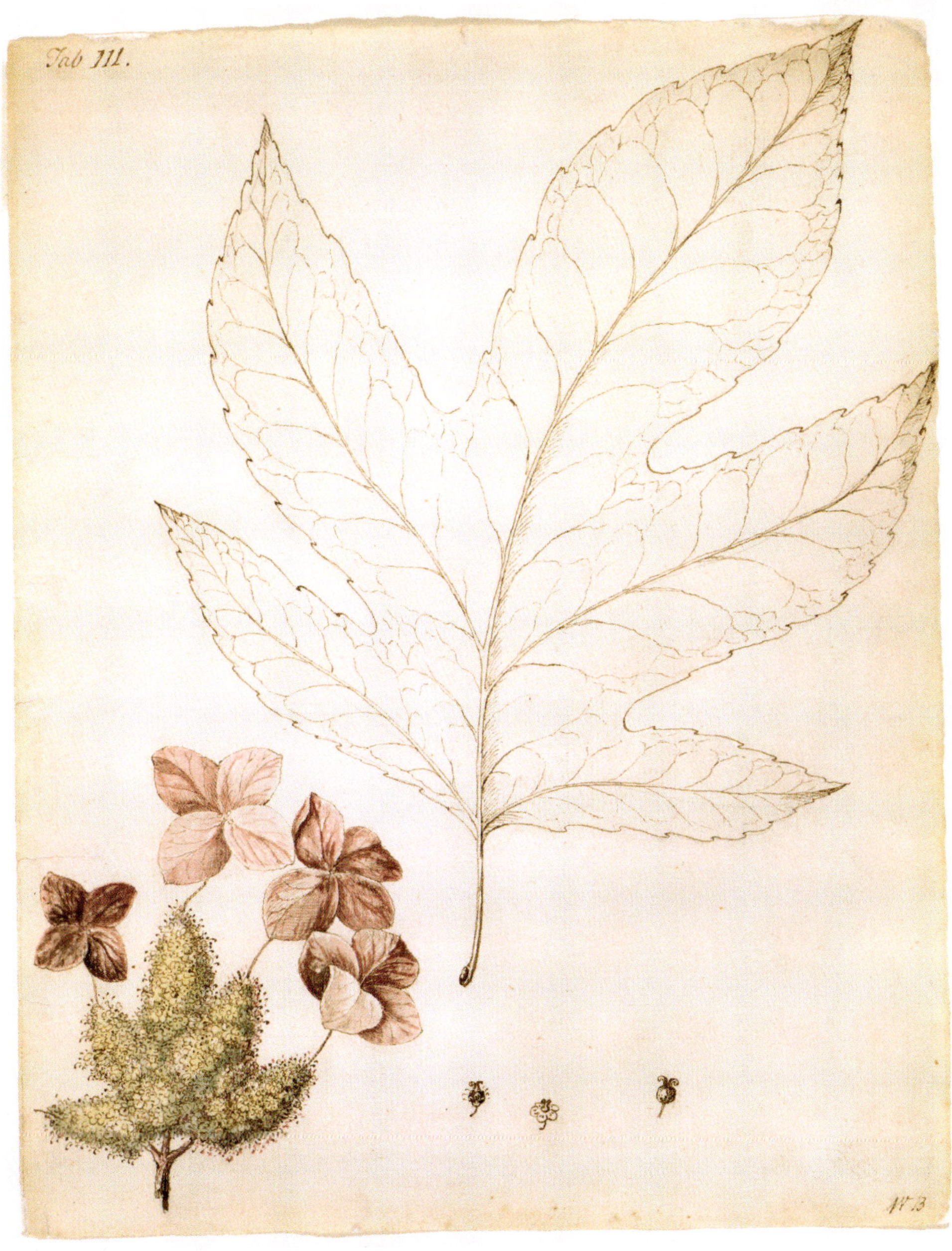

PLATE 9. William Bartram, *Hydrangea quercifolia* [Oakleaf hydrangea], 1788. © Natural History Museum, London.

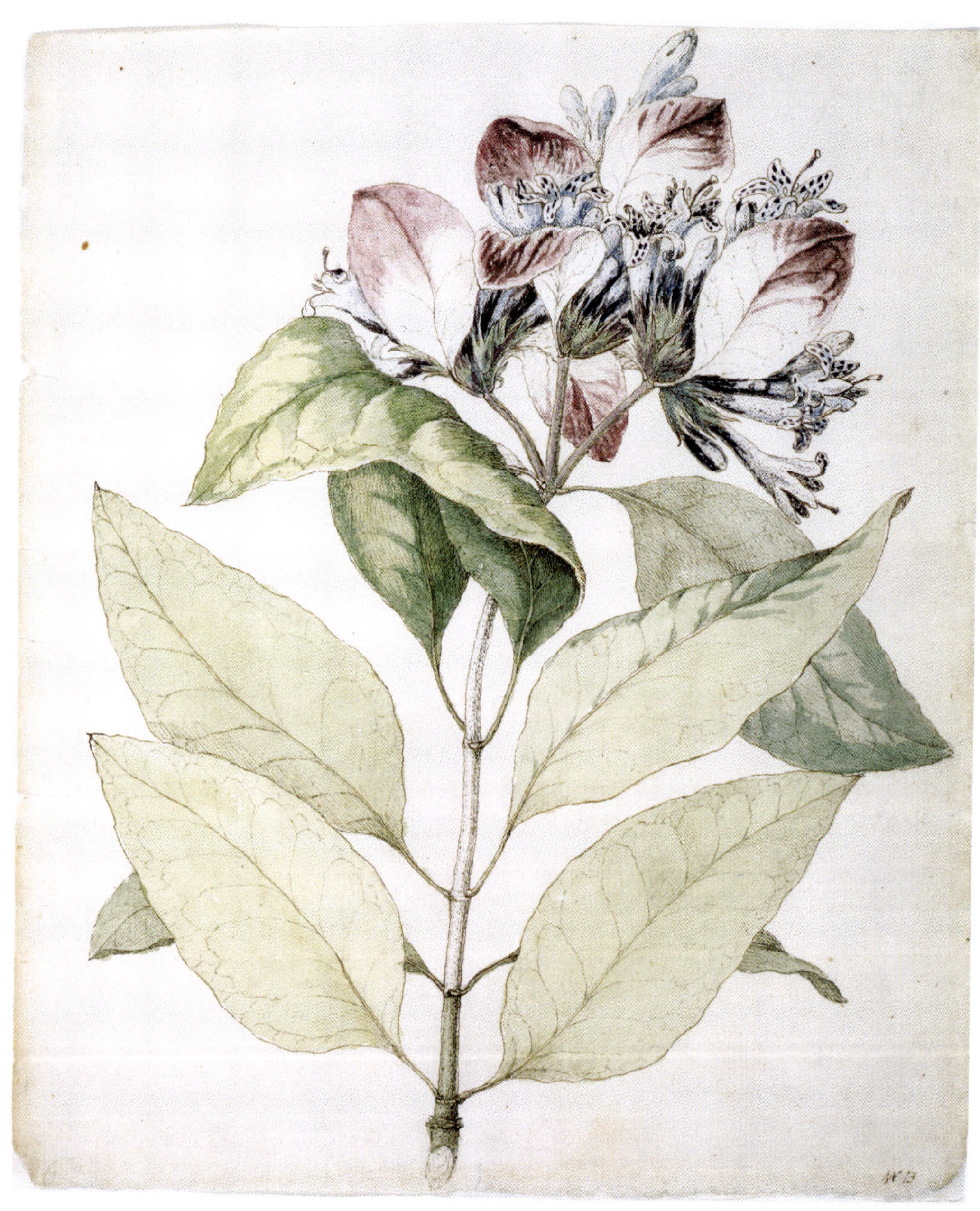

PLATE 10. William Bartram, *Anonymos. Bignonia bracteata* [*Pinckneya bracteata* or Georgia feverbark], 1788. © Natural History Museum, London.

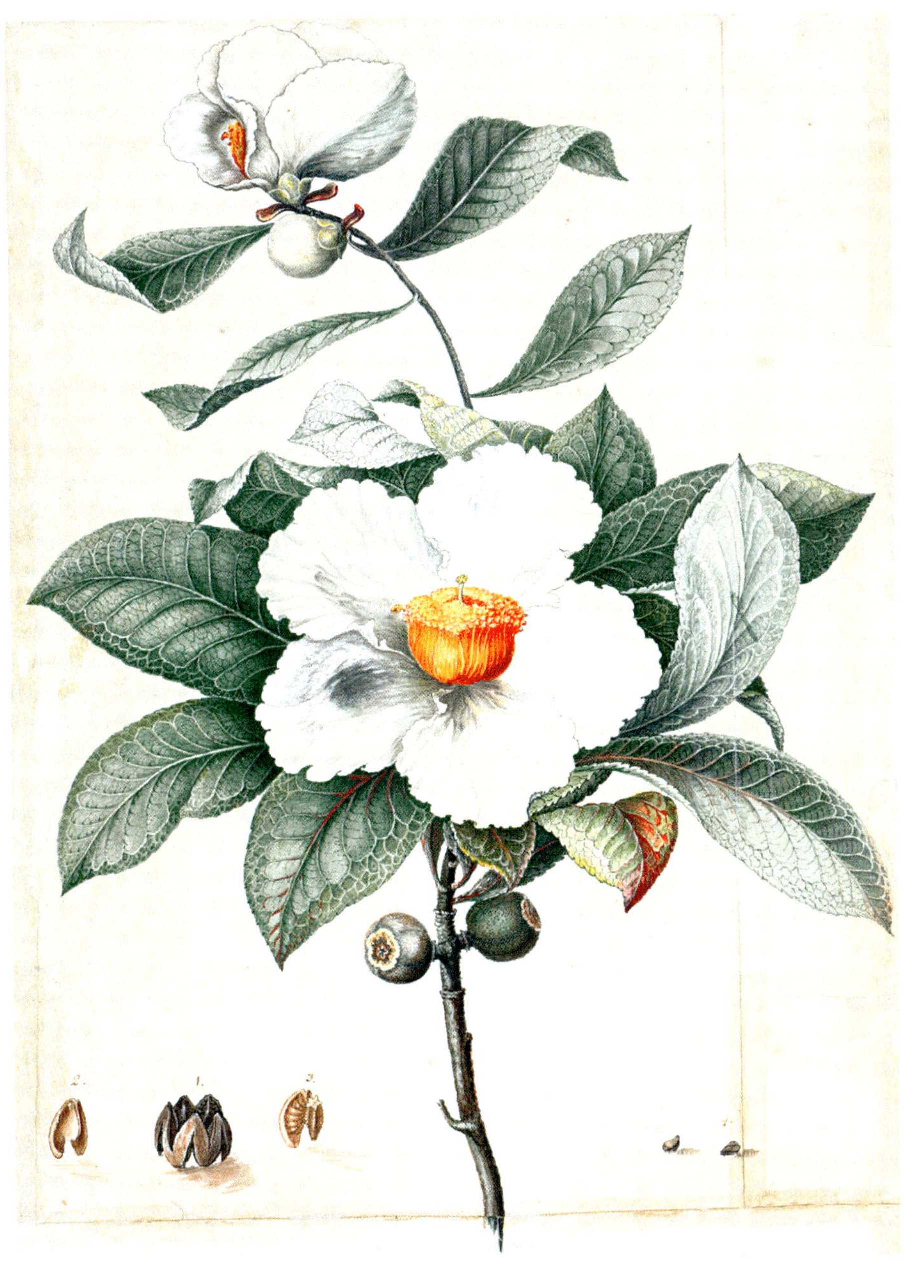

PLATE 11. William Bartram, *Franklinia alatamaha* [Franklin tree], 1788. © Natural History Museum, London.

Chapter 4

A World in Figures

Bartram made clear in his natural history that he saw no external point from which to apprehend nature. Moreover, he understood the natural world as ever in flux, perpetually shifting in accordance with a divinely ordained, if indiscernible, plan. The mutability of the cosmos and his own embeddedness in it yielded a representational dilemma. How might one portray a dynamic natural world of which one is an inextricable part? Rather than try to capture an organism's true form—something stable and irreducible, but also inert—Bartram directed his attention toward conveying its liveliness and interactive qualities. Yet this emphasis came with problems of its own— namely, verifying for his audience that his representations were accurate and not mere flights of fancy.

The particular representational struggles Bartram faced call to mind the phrase *ad vivum* or "from the life," a designation that Claudia Swan and others have examined in detail.[1] The phrase has been used by artists to authenticate visual representation in a number of ways, not necessarily mutually exclusive. Drawing from the life can indicate that the image exudes a certain liveliness that elicits an engaged response from the viewer, as though the representation itself were alive; it can also indicate that the artist is a trustworthy observer, faithfully delineating a tangible object he or she has seen firsthand. Perhaps more to the point is how it can, in the words of Robert Felfe, "mark a specific understanding of art as an uninterrupted continuation of natural effects"—that is, that the naturalist's representation of the world is not separate from but a part of nature, continuous with its relational patterns and temporal processes.[2] In his natural history images, Bartram mustered all of these elements to reinforce the truth and accuracy of his observations.

The Animating Figure

In Bartram's natural history drawings, a sense of liveliness is of preeminent importance. The mutability of nature forecloses the possibility of any true form, and any representation purporting to convey it would ultimately be misleading. Images that elicit active engagement from the viewer, however, reinforce a sense of liveliness and vitality that can be more lifelike than strict verisimilitude. In her work on "living presence response" in the early modern period, Caroline van Eck connects these two opposing conceptions of visual representation to episodes in Ovid's *Metamorphoses*. Representation driven by a mimetic imperative derives from encounters with the Gorgon Medusa, who literally petrified living beings and turned them into stone. Representation aimed at liveliness and vitality traces its origin to Galatea's doting attention and skillful craftsmanship, which turned his statue Pygmalion into flesh. Representation, van Eck points out, can deaden what it portrays, or it can enliven it.[3]

Bartram countered the potential petrifying effects of representation through his compositional strategies, such as the accretion of perspectives, yet his most important corrective may have been his choice of medium. In examining the medium of drawing specifically, David Rosand emphasizes its inability to conceal the conditions and gestures of its making, allowing the viewer a certain kind of access to the artist's mental and manual processes. Because of this, drawing gains authenticity through its direct relationship to the artist.[4] The inscribed line was believed to refer not only to the object represented but also, as the record of a manual gesture, back to the draughtsperson, highlighting the image maker as the hinge between image and represented object.[5]

The ambivalence of linear representation, referring both out toward the world and back toward the artist, creates the possibility for a more dynamic form of portrayal, one whose meaning is mimetic but also indexical.[6] Its operation in multiple representational registers demands the viewer's participation and imaginative engagement to interpret it, evoking the pleasures of pursuit so central to William Hogarth's aesthetics. And it is this imaginative collaboration of artist and viewer—in effect, their joint re-creation of the world—that invests Bartram's representations with life. In his desire to convey the vitality of nature and his participation in it, he deployed a complex graphic language as dynamic as nature itself.

Bartram would undoubtedly have been familiar with Pliny the Elder's *Naturalis Historia* and his discussion of Apelles, who was considered an exceptionally gifted artist because of the delicacy of his outlines. To illustrate Apelles's mastery, Pliny related an anecdote in which the artist left "an outline of singularly minute fineness" as his calling card at the studio of his colleague Protogenes. Returning to his studio, Protogenes "instantly exclaimed that Apelles must have been the visitor, for no other person was capable of executing anything so exquisitely perfect."[7] Pliny's story emphasizes the representational playfulness of the drawn line, which can refer to the artist as well as to the thing represented.

In *The Analysis of Beauty*, Hogarth would interpret this anecdote as the foundational story for his own taxonomy of lines. "If it was only a stroke (tho' as fine as a hair as Pliny seems to think) it could not possibly . . . denote the abilities of a great painter," he maintained. He suggested instead that Apelles drew not just an incredibly delicate line but an especially expressive one—what Hogarth would call the serpentine line or the Line of Grace.[8] He described the serpentine line as a compound of straight and curved strokes that yields a more informative and beautiful mark. Alone, straight and curved lines vary only in length and degree of curvature, but when joined together they articulate an unlimited variety of forms and surfaces. The serpentine line's curves and coils lead the eye in active pursuit, helping the viewer to envision the depicted object's liveliness.[9] Both Pliny's history and Hogarth's linear system present line as the "graphic seam" between sensory experience and intellectual construct. In Bartram's natural history drawings, line delineates nature as he experienced it while also encouraging the beholder to retrace visually his movements and gestures.[10] The self-referentiality of his drawings authenticates Bartram's depictions, just as their looping skeins, dashes, and stipples capture the dynamism of the natural world.

William Bartram's early investigations of the expressive possibilities of line appear in an unlikely place: as an inscription inside a copy of Carl Linnaeus's *Genera Plantarum* (2d ed., 1742). The book was originally a gift to John Bartram from the Dutch botanist Jan Frederik Gronovius, who provided natural history texts in exchange for specimens of American plants. Beyond *Genera Plantarum*, which John received in 1743, Gronovius sent editions of Linnaeus's *Systema Naturae* (4th ed., 1744), *Fauna Suecica* (1746), and *Bibliotheca Botanica* (2d ed., 1751), as well as his own publications, *Flora Virginica* (1739) and *Index Supellectilis Lapidae* (1740).[11] These books were enormously valuable, since they could be difficult to come by in Philadelphia. Local bookstores did not generally stock works of natural history and natural philosophy, and as late as 1746 the Library Company held only sixty related titles.[12] John attributed much of his progress as a naturalist to his European contacts and the materials they provided. Yet even as he worked toward mastering the books' contents, he complained that they often lacked important details or material observations, indicating the inherent incompatibility between the natural world and the rationalizing structures designed to order it.

William, too, would come to articulate this incompatibility in his drawings, as he forged his own form of representation. Although most of his drawings from the 1750s echo the etched line and invented compositions of his models and mentors, Mark Catesby and George Edwards, the gifted copy of *Genera Plantarum* offers a clear sense of his early experimentation. When John passed the book to his son in June 1755, William noted its transfer on the flyleaf, where he also drew a decorative figure (fig. 4.1). This unusual emblem appears the production of distracted thought or dreamy reverie, and yet—as the most compelling visual element of William's inscription—it demands attention. Reverie is not synonymous with mere mindlessness but is instead more aptly conceived as a moment of pre-imaginative possibility, as open ground in which ideas and concepts may be sown.[13] Considered in this light, William's inscription reads not as distraction but as his early *figuring out* of a theory of representation. As this and later drawings attest, the expressive capabilities of line would play a key role in that theory's development.

The emblem's form and placement initially evoke a bookplate, akin to the heraldic design pasted in the Bartram family bible (fig. 4.2). J. Hector St. John de Crèvecoeur described a framed version of this coat of arms in his account of the "Bertram" home and garden in *Letters from an American Farmer* (1782).

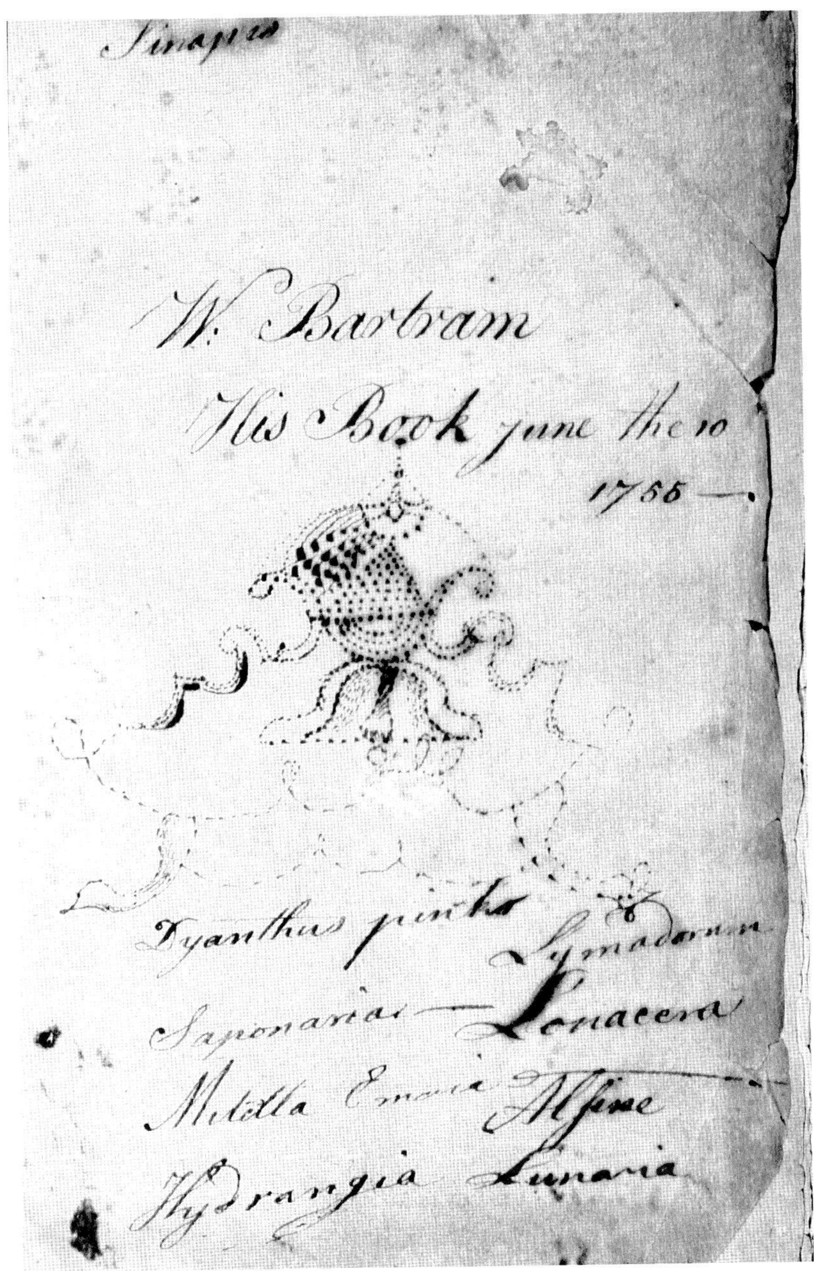

FIG. 4.1. William Bartram, inscription in his personal copy of Carl Linnaeus, *Genera Plantarum*, 2d ed. (Leiden, 1742), 1755. Bartram's Garden, Philadelphia.

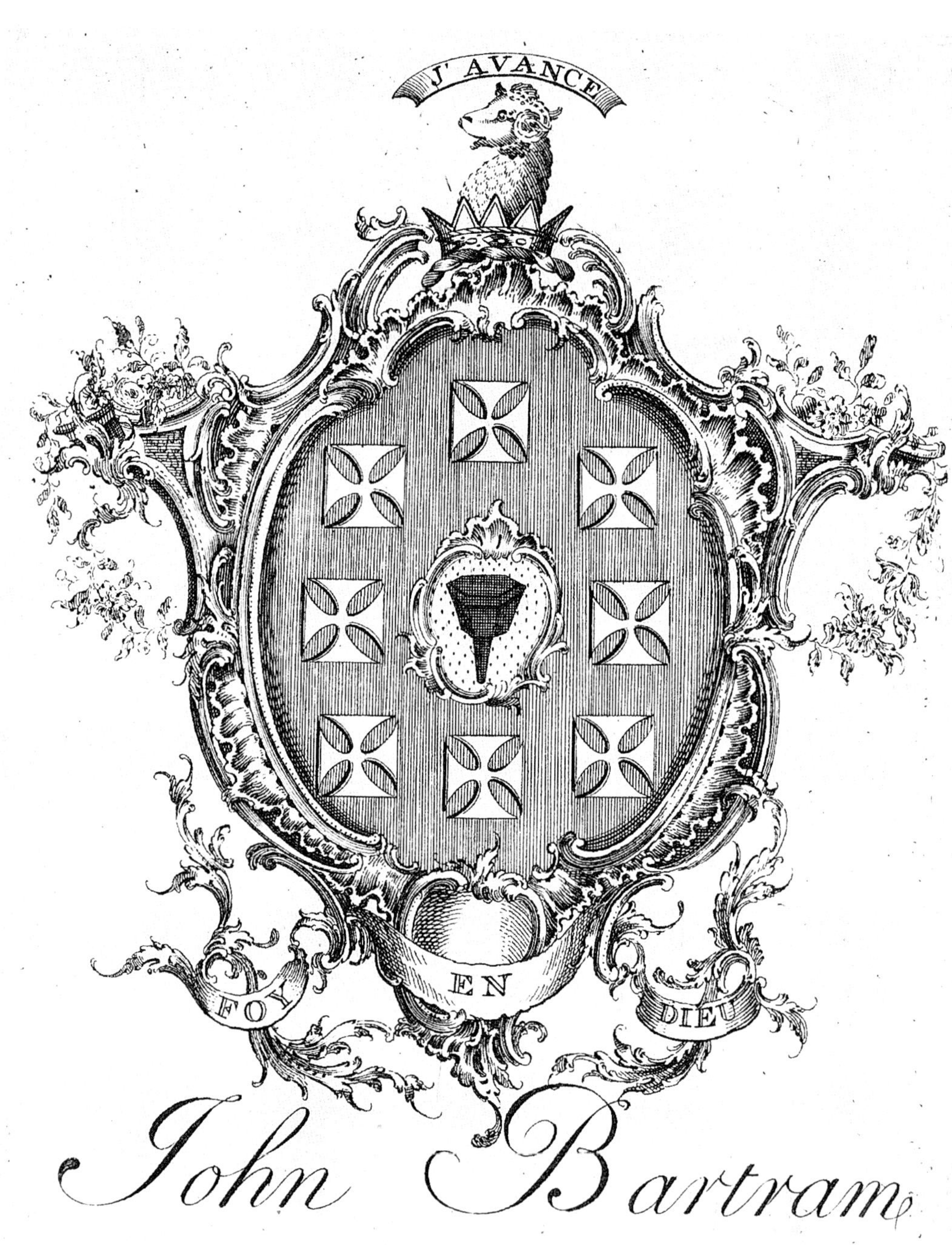

Fig. 4.2. John Bartram bookplate, after 1760. Courtesy, American Antiquarian Society.

"My father was a Frenchman," Crèvecoeur's semifictional John Bartram claims, "he brought this piece of painting over with him; I keep it as a piece of family furniture, and as a memorial of his removal hither."[14] The Bartrams were not of French descent, nor was John's father an émigré who carried with him this coat of arms, which was likely designed in Philadelphia sometime about 1760. It borrows the ram, crown, and eight crosses from a Scottish branch of the Bartram family but includes in the center a stonemason's mallet—likely a reference to the stone house John built at his garden and to the cider press he carved into the rock of the Schuylkill riverbank.[15]

Unlike the ram, crown, or mallet in his father's invented coat of arms, William's emblem yields no clear symbolic or mimetic relationship to the visible world; instead, it offers an odd concatenation of lines and stippled marks that almost but never quite coalesce into identifiable form. Its central figure is composed of a half circle and a skewed triangle, which are balanced atop three curved shapes that recall the base of a tilt-top table. It is here that William came closest to representing an object observed from life, with a modicum of consideration paid to conventions of perspective. The middle foot, for example, is shaded and foreshortened, in accordance with the *S*-curve seen in the feet in profile yet concealed in this frontal view. The stippled patterns of the half circle and triangle, however, deflate the fledgling volumetric quality and draw attention to their flat, slightly cockeyed geometry. The decorative elements that sprout from and encircle the central figure contribute to the emblem's flatness, pointing to the dissolution not simply of volume but of form entirely.

Rosand observes that the drawn line always resists full capitulation to a mimetic imperative, and such resistance is particularly heightened here. Bartram's waving lines and stipples play along the edges of pictorial representation—suggesting at points a tureen, table legs, even a ship's rigging—while resolutely declaring their own presence as handmade marks.[16] The pooling stains that are drawn into whisper-thin lines underscore the fluidity of the ink and the absorbency of the paper. The marks are less a form of representation than one of self-presentation, as they call attention to their materiality and to the artist's gestures in their creation. Hogarth observed that a graceful line could only be the product of a graceful gesture, "insomuch that the hand takes a lively movement in making it with pen or pencil."[17] As with Apelles, the line always refers to the creator and that person's bodily movements, regardless of its other representational functions.

In this sense, Bartram's emblem is as much his signature as the name above it. The textual inscription also marks out his movement, but it is movement that is contained and controlled by the letters it must form.[18] The decorative figure, on the other hand, is line sprung loose from the dictates of text or image. Instead, it is Bartram's exploration of the full possibilities of ink, quill, and paper, and his marks range from thick and full to thin and skating. Although they sometimes cohere into near legible form, as in the ornament's base, they also consistently undermine this emergent legibility, as in the irregular, bannerlike shapes that cascade from its sides. These figures suggest decorative embellishment, but at same time they seem to embellish nothing—ornaments without an object to adorn or, perhaps, ornaments of an ornament. As these scrolls unfurl and their stippled lines become lighter and sparer, they lose their association with embellishment and suggest pure movement through space.

As records of the artist's manual gestures, these marks and lines also spur the imagined movement of the observer. In his *Analysis*, Hogarth recalled being so bewitched by the comportment of a favorite dancer that his eye "was dancing with her all the time." The sinuousness of waving and coiling lines are similarly bewitching to the viewer, he argued, leading *"the eye [on] a wanton kind of chace,"* as the observer empathically traces the artist's gestures.[19] The dynamic perception encouraged through compositional means—the cobbling together of perspectives in the plates for *The Analysis of Beauty*, say, or in Bartram's drawing of the American lotus for Collinson—can also be activated by a single mark: the serpentine line. Although Bartram's decorative figure is not mimetic, it still represents movement through its loping scrolls, at once capturing the artist's movements and, through the imaginative engagement of the viewer, conveying the very idea of motion.

Bartram's inscription is not usually incorporated into his broader oeuvre, even though its serpentine lines recur throughout his mature work. These lines are found in the snaking rivulets of his Alachua Savanna maps and the spiraling tendrils of his balsam pear, yet it is in his 1774 drawing of the coachwhip snake (*Masticophis flagellum;* fig. 4.3) that their capacity to animate is most precisely balanced against visual likeness. One of about thirty-eight works sent to John Fothergill during Bartram's southern travels, the drawing depicts a species of snake indigenous to Florida that was known not for its predatory behavior but for its swiftness. It conveys the snake's appearance—an excerpted detail highlights variations in the markings of its head—while the sinuous wreaths of its body encourage the viewer to reexperience its movements.

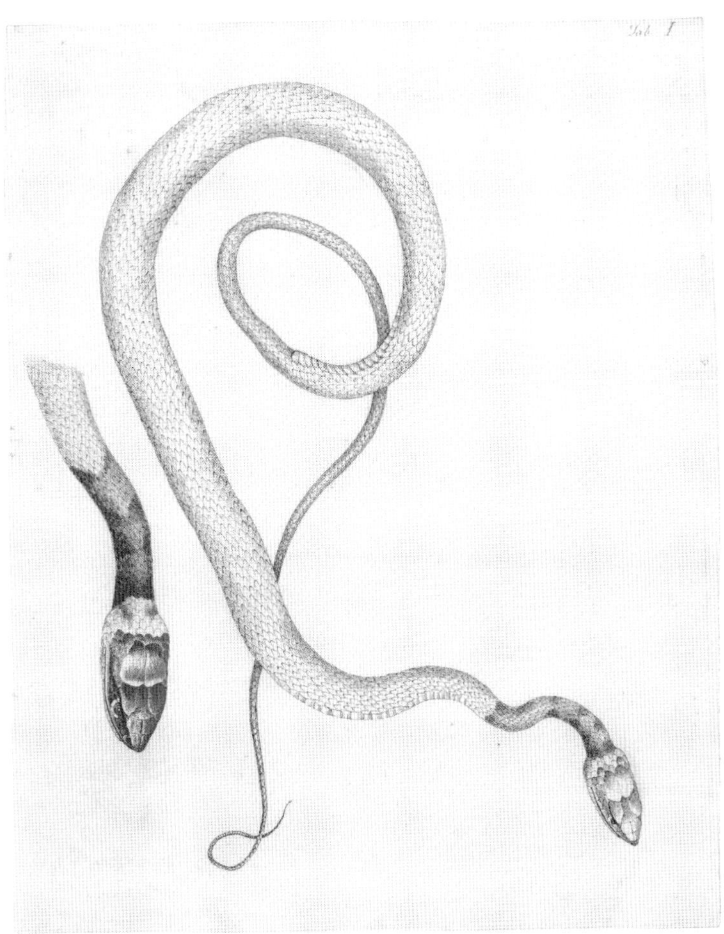

Fig. 4.3. William Bartram, *Coach Whip Snake* [*Masticophis flagellum*] *from Et Florida*, 1774. © Natural History Museum, London.

American snakes were of great interest in Europe, and they were considered emblematic of the new country's mysteriousness. The pages of the Royal Society's *Philosophical Transactions* teem with descriptions of snakes, serpents, and vipers in the New World, though the Fellows paid special attention to the rattlesnake, which was invested with almost supernatural powers.[20] The rattlesnake was believed to "fascinate" or mesmerize its prey—not unlike the serpentine line's ability to captivate—and to transform its human victims into snakelike creatures. Joseph Breintnall's account of a rattlesnake bite found its way into the society's *Transactions* through Peter Collinson. "[My Arm] seem'd almost void of Feeling," Breintnall observed, "yet it would work, jump, writhe, and twist like a Snake in the Skin, and change Colours, and be spotted."[21] Still, other American snakes came with their own myths. Some believed that the glass snake could reattach its fractured parts (perhaps a model for Benjamin Franklin's *Join, or Die* illustration), while others thought the coachwhip snake could cut a man in two with a snap of its tail.[22] The mythologies surrounding American snakes underscored the perception of the country as a veritable Garden of Eden: beautiful, yet fraught with potential danger.

Throughout Bartram's natural history, his descriptions of snakes focus less on their mythical qualities than on their appearance and movements. The rattlesnake, though "very beautiful, speckled and clouded," is slower than a child can walk, and the glass snake—so named because it appears to be made of blue-green glass—is only slightly more nimble.[23] The coachwhip snake, however, is not only beautiful but dexterous and fleet. Bartram claimed to have seen one tightly coiled around a hawk, attempting to protect itself from predation. When the hawk finally struggled free and the coachwhip snake slid away, Bartram was surprised by its speed. "They are extremely swift, seeming almost to fly over the surface of the ground," he wrote, "one very fine one accompanied me along the road side . . . although I proceeded on a good round trot on purpose to observe how fast they could proceed."[24] His attention to its undulating motions suggests the coachwhip snake served as a natural analogue of Hogarth's serpentine line.

With his drawing of the snake, Bartram articulated the dynamic potential of linear representation. At its most basic level, the drawing is but two calligraphic marks—a short nearly straight line and a long sinuous one composed of three coils and a rolling S-curve.[25] Bartram cleverly played with the meaning-making power of these calligraphic marks, using them to orient and disorient the viewer. Someone who reads left to right, as did all of his patrons, would inevitably orient the drawing horizontally, interpreting the full snake as weaving toward the upper right corner of the page.[26] A small, easily overlooked notation, however, forces a rotation of ninety degrees clockwise, and what had been the left edge of the drawing becomes the top, with its finely penciled "Tab. I." The depicted snake, consequently, begins to gain speed as it slides downward rather than up (plate 6). In encouraging the movement of the viewer, who rotates or walks round the drawing, these lines and notation reinforce the snake's alacrity.

The sparseness of Bartram's coachwhip snake reads not unlike Hogarth's schematic depictions of country dancers in the upper left corner of plate 2 in the

Analysis, which are intended "to shew how few lines are necessary to express . . . different attitudes."[27] At the same time, Hogarth observed, "every figure is rather a suspended action . . . than an attitude," and he suggested the best way to convey the gracefulness of a dance was to trace the dancers' movements in space. He complemented these schematic depictions with diagrams of the hay and "its cypher of S's," and the minuet, which contains as "many movements in the serpentine lines as can well be put together."[28] Yet in his diagrams of the dances' figures, the dancers themselves are effaced (see fig. 2.14). Hogarth's faith in linear representation meets here a seemingly insoluble problem: a single line can convey either an object or its movement through space but never both. Bartram attempted to cut this Gordian knot by fusing Hogarth's two schematic renderings, with the line of the snake's body delineating its corporeal attitude as well as the history of its movement.

Bartram's coachwhip snake brilliantly offers an antidote to the potentially petrifying effects of representation based on verisimilitude. Here the naturalist translated his sensory experience of the snake's movement into a gestural one, through the serpentine line. Although this calligraphic mark re-creates the creature's movement by encouraging—even demanding—the viewer's own, either actual or imagined, it also approximates the snake's visual appearance. This form of dynamic representation traces back to the linear experimentation on the flyleaf of *Genera Plantarum.* Its spooling central figure indicates not only his ownership of the book but perhaps also an earnest attempt to animate its lifeless lists. In his cluster of stipples and unfurling lines Bartram captured nature's vitality, even though the emblem bears no clear mimetic relationship to the natural world. The flyleaf points to Bartram's own method of representation, one that emphasizes the energy, movement, and liveliness of nature as much as its appearance. His drawings often gain their lifelikeness not simply from a visual approximation of a depicted object but from the engagement they demand of the viewer.

Figure and Ground

Bartram's use of figures as a means of animating the natural world was not limited to his graphic work but was likewise incorporated into his textual descriptions. As a student of classical rhetoric, he would have been aware that figures were essential to enlivening both visual and verbal descriptions. As van

Eck notes, the Latin term *figura* suggests something that shifts from one form to another, analogous one; the "squadrons" of savanna cranes Bartram observed in the Alachua Savanna are one such example. The sense of movement implicit in the term, the shift from one meaning or form to another, is what makes figuring such a powerful rhetorical device. In classical rhetoric, van Eck explains, *figura* can refer "to all expressions that have 'received a new aspect.'"[29] The purpose of these various rhetorical turns is to encourage the observer or auditor to see for themselves—that is, to insert themselves in the scene or episode described. Indeed, it is this notion of *figura* that Hogarth so embraced in his *Analysis*, encouraging observers "to see *with our own eyes*."[30]

As much as the use of figures was essential to both Bartram's writings and images, he deployed them somewhat differently in the two arenas. His text, for instance, is stuffed to bursting in a way that his drawings are not. The language in *Travels* is richly embroidered, occasionally exhausting, struck through with a "wondrous kind of floundering eloquence," as it shuttles among his lavish botanical and zoological descriptions, philosophical ruminations, and exclamations of God's beneficence.[31] The poetic effusions and rich texture of his writing spin a dazzling world of natural drama.

In his *Travels*, for instance, Bartram regaled the reader with a dramatic tale of the "subtle, greedy alegator," which took place along the banks of the St. Johns River in East Florida.[32] His extended description is terrifying, its beginning transposed into the present tense, as though the alarming experience was ever near: "Behold him rushing forth from the flags and reeds. His enormous body swells. His plaited tail brandished high, floats upon the lake. The waters like a cataract descend from his opening jaws. Clouds of smoke issue from his dilated nostrils." Bartram's fear for his own safety tracks through the following pages, as he described being surrounded by numerous alligators that "belch[ed] floods of water" over him and stole his provisions. When he finally discovered the reason behind their sheer numbers, it "exceeded every thing of the kind I had ever heard of."[33] His rational processing of this information demands the passage be quoted at length:

> How shall I express myself so as to convey an adequate idea of it to the reader, and at the same time avoid raising suspicions of my want of veracity[?] Should I say, that the river (in this place) from shore to shore, and perhaps near half a mile above and below me, appeared to be one solid bank of fish, . . . and that the alligators were in

such incredible numbers, and so close together from shore to shore, that it would have been easy to have walked across on their heads, had the animals been harmless. What expressions can sufficiently declare the shocking scene that for some minutes continued, whilst this mighty army of fish were forcing the pass? During this attempt, thousands, I may say hundreds of thousands of them were caught and swallowed by the devouring alligators. I have seen an alligator take up out of the water several great fish at a time, and just squeeze them betwixt his jaws, while the tails of the great trout flapped about his eyes and lips, ere he had swallowed them. The horrid noise of their closing jaws, their plunging amidst the broken banks of fish, and rising with their prey some feet upright above the water, the floods of water and blood rushing out of their mouths, and the clouds of vapour issuing from their wide nostrils, were truly frightful.[34]

Bartram's observations of alligators challenged even his capacity to believe, forcing him to interrogate his own senses. This detailed passage deploys tense and richly detailed description and figurative language to make the event unnervingly present to the reader. Bartram relied on the classical rhetorical device of *enargeia* or "vividness," in which words are used to trigger the readers' memories—their storehouses of mental images—that allow them to re-create and experience the scene.

Threaded through this passage, however, are also Bartram's reflections on the inadequacy of all linguistic devices to convey the experience, demanding he turn to visual representation. In his drawings he eschewed the descriptive density of his text, taking a somewhat different approach. In *The Alegator of St. Johns* (ca. 1774; fig. 4.4), for example, he drew a spare landscape border that contrasts with the detailed visual description of the reptiles at its center, and his *View of the Alegator Hole* (ca. 1774; fig. 4.5) exhibits an unexpected drop off in tone and detail toward the drawing's edges, as though the world Bartram portrayed emerged from the dark mass at its center.[35] In his visual renderings of the scene, he relied heavily on an interplay between emptiness and plenitude that demands viewers' imaginative completion.

Michael Gaudio analyzes the appearance of such voids in Bartram's drawings, seeing them as the naturalist's recognition of epistemological limits, of gaps in knowledge that cannot be filled and thresholds that cannot be crossed.[36] Bartram was clearly fascinated by the limitations of what could be perceived or definitively known, and he often described objects and sites that challenge

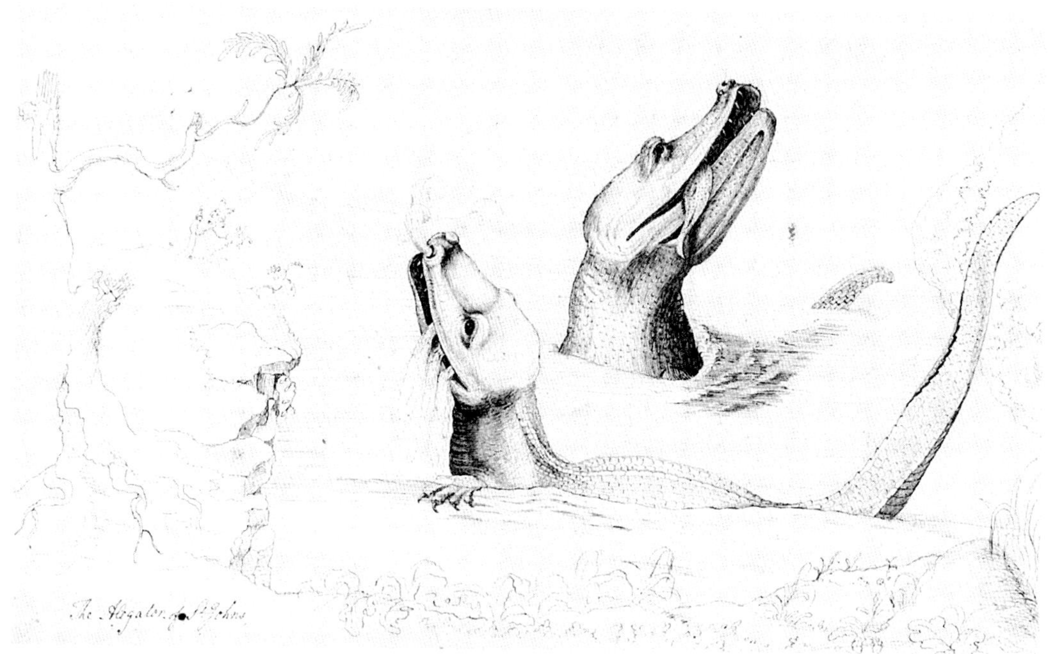

FIG. 4.4. William Bartram, *The Alegator of St. Johns*, 1774–1775. © Natural History Museum, London.

one's powers of observation and discernment in reverent terms. For Bartram, the sinkholes along the edges of the Alachua Savanna were "very wonderful work[s] of nature" that, he could only presume, led to "secret subterranean conduits and gloomy vaults, to other distant lakes and rivers."[37] He was captivated by how natural productions eluded perceptual and cognitive mastery; in their unknowability, they left space for the imagination. By re-creating such voids in his drawings, Bartram encouraged his audience to engage in similar cognitive processes.

In classical rhetoric, memory and imagination are essential to creating a living presence response. As van Eck observes, the description triggers a recall of memories that the imagination then reconfigures before the mind's eye.[38] Such a model of the imagination was popularized in the eighteenth century through the work of the essayist Joseph Addison. According to Addison, imagination works in a recombinative mode, drawing on sensory impressions that had previously been amassed but then can be endlessly refigured. The imagination possesses the "power of retaining, altering, and compounding those [sensations], which we

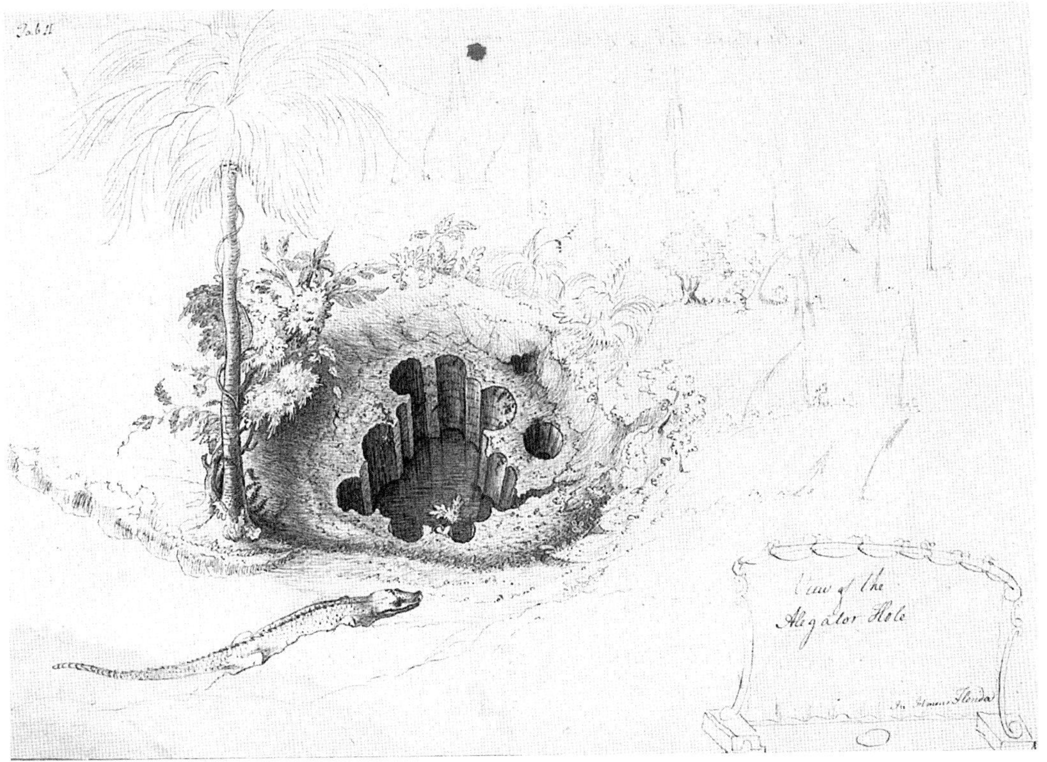

FIG. 4.5. William Bartram, *View of the Alegator Hole*, 1774–1775. © Natural History Museum, London.

have once received, into all the varieties of picture and vision," with the potential to bring a full world into being.[39]

Although in *Travels*, Bartram endeavored to create a powerful sense of vividness, it is in his drawings that he most urgently demanded the audience's imagination, through his deployment of presence and absence, complex figures and empty grounds. His graphic work allows for greater ambiguity, thus requiring a higher degree of imaginative work and engagement than his textual descriptions of the same events.[40] His *Alegator of St. Johns*, for instance, depicts two of the feeding alligators at its center, with "the floods of water and blood rushing out of their mouths, and the clouds of vapour issuing from their wide nostrils."[41] Their heads, necks, and tails are heavily detailed with series of checkmarks to denote their scaly exteriors, and the rest of their bodies are submerged in a strangely viscid substance. The thickness of this central passage contrasts with the thinly rendered landscape that surrounds it, where delicate lines suggest low-lying plants and a truncated tree atop a rocky embankment. The water-belching, smoke-blowing alligators are the primary subjects of the drawing, but the most graphically compelling passage occurs along the left border, where the embankment's contours transform into a ragged tree trunk.

Composed of three primary lines that shift in meaning as they skate across the paper's surface, the tree and the embankment push the ambivalence of the drawn line to its limits. The outline of the tree, beginning about a quarter of the way down the left edge of the paper, traces the contours of its leafed branch, the flat edge of an amputated limb, and the length of the trunk, before transforming into a figure that evokes a crevice in the embankment. Similarly, the tree's underground root system is structured around two lines that slowly snake toward each other, forming an inverted *V* that becomes a hollow in the trunk.[42] These lines delineate a visible absence above ground (the hollow) and an invisible presence below (the life-sustaining roots), actively punning on questions of visibility and invisibility. They scramble distinctions between figure and ground, above and below, forcing the viewer into a series of imaginative recombinations.

Only occasionally in the *Travels* did Bartram punctuate his verbal plenitude with such voids. These occur in the form of *nomina nuda*, names he applied to natural objects that appear to be accepted taxonomic designations. Although naturalists, then as now, have the prerogative of naming their discoveries in print, for them to be accepted the names must be accompanied by full taxonomic descriptions. In *Travels*, Bartram often introduced terms with no accompanying taxonomic description or, in fact, any description at all. These *nomina nuda* do not necessarily apply to referents in the world but, rather, seem to function as empty symbols that enact absence. The reader's attempts to yoke them to referents in the real world are merely misguided efforts to fix figure and ground in Bartram's natural history. This ultimately proves a profoundly challenging task since, as Christopher Iannini explains, nature functions both as parable and as the ground literally traversed by Bartram.[43]

Obscurity and absence are again taken up in Bartram's *View of the Alegator Hole*, where they are conveyed symbolically as well as graphically. In his *Travels*, he described the site as "a very curious place, . . . it is one of those vast circular sinks, which we behold almost every where about us as we traversed these forests, after we left the Alachua savanna. . . . [A] very large alligator at present is lord or chief; many have been killed here, but the throne is never long vacant."

The drawing features a scalloped sinkhole at its center, partially filled with water that is neither transparent nor reflective. Like the fluid in *The Alegator of St. Johns*, it is murky, gelatinous. Foliage surrounds the depression, and a tall palm stands on the left, with a thin vine coiled around its trunk. The level of information provided in the drawing diminishes as we move out from the center: the sinkhole is thickly described, the palm fainter and more stylized, and the surrounding savanna barely suggested. An alligator—the sinkhole's "lord and chief"—slices through the blank foreground, and a cartouche in the lower right corner offers the drawing's title.[44]

The drawing's foreground elements of alligator and cartouche operate in its symbolic register. Although the alligator would seem to function mimetically (it is, after all, the alligator hole), the reptile was also a symbol of plenitude. Its inclusion in a natural history cabinet typically signaled the collection's richness and diversity because of the difficulty of capturing and transporting the creature.[45] Hans Sloane even attempted to bring a live alligator to England after his fifteen-month sojourn in Jamaica, but the reptile died in transport.[46] As a symbol of excess, the alligator contrasts with the curtained cartouche in the composition's lower right corner, a signifier of emptiness or blankness. The curtain that bears the drawing's title is only minimally described; hatches indicate a slight undulation of the fabric, and a series of jittering loops suggests the strings that fasten it to a horizontal C-scroll.

An unusual addition to the drawing, the drape is as significant as it is curious. Like the serpentine line that recalls Apelles's linear "signature," the drape likewise recalls a foundational story of representation as described by Pliny the Elder. It summons the spirited contest between the classical Greek painters Parrhasius and Zeuxis, in which Parrhasius defeated his colleague with a lifelike portrayal of a curtain. Zeuxis had painted a cluster of grapes so believable that it duped a flock of birds, but Parrhasius's painted drape fooled Zeuxis, who demanded it "be drawn aside to let the picture be seen."[47]

This anecdote has long been recognized as articulating representation's mimetic imperative, but it also describes how art can create a living presence response in the viewer.[48] The competition between Parrhasius and Zeuxis aptly illustrates the role of presence, absence, and imagination in creating a lifelike image, since the curtain requires the viewer to imagine what it conceals. The particular interplay of presence and absence described in the story of Zeuxis and Parrahasius is recapitulated in the graphic structure of Bartram's *View of the*

Alegator Hole, in which the area of greatest visual information—the sinkhole—is also its most obscured. The sinkhole's projections, recesses, and shadows are all finely delineated, but the pool itself is opaque. The water neither reflects its surroundings nor reveals its contents, demanding the viewer fill in what is so enticingly withheld. In obscuring the contents of the sinkhole, Bartram transformed the viewer from receiver to participant, whose imaginative engagement with the scene invests it with the vitality that verisimilitude alone fails to capture.

Authenticating Monstrosity

Exotic though they were, stuffed American alligators and snake skins were objects that European virtuosi might very well have access to in natural history collections, allowing them to compare Bartram's drawings to the prepared specimens themselves. Yet some natural objects, unable to withstand an uncertain ocean voyage, resisted such preparation and transport. The absence of an accompanying specimen was a point of special concern for artist naturalists, since they inevitably had to portray objects unlike anything previously known, objects that might seem invented or fantastical.

The conceit of "from the life" could mean making the drawing lively, in that it engenders a living presence response in the viewer, or it could mean reassuring viewers that the representation portrays an existing object in the world. One method of making such images read as lifelike—in the truthful, rather than animated sense of the term—was the use of familiarizing strategies, such as the construction of tableaux or the deployment of recognizable visual idioms. George Edwards's invented environments, with their deciduous trees, smatterings of pebbles, and tufts of grass, invariably suggest a climate that would have been comfortably familiar to his readers, even when these environments host flora and fauna from equatorial regions. Robert Hooke's and Henry Baker's microscopic views similarly transformed the bewilderment of magnification into scaled-down but recognizable landscapes, where blight becomes undulating fields or a frog's veins transform into a beautiful network of "Rivers, Streams, and Rills."[49] Bartram certainly cited some of these strategies, but it was almost always to undermine their capacity to familiarize. In his drawing of the American lotus, for instance, he quoted Edwards's placement of a heron at the edge of a pond but entirely upended the tableau's sense of scale to make it more, not less, radically strange.

Bartram possibly found such familiarizing tactics too prescriptive, foreclosing the aesthetic pleasure of pursuit and discovery. They dampened the wonder of the marvelous specimen in order to make it fit within the viewer's preconceptions. To authenticate his drawings, Bartram instead emphasized the relationship between his observations and his process of figuring. He followed the model of the seventeenth-century botanist John Banister, who had explored North American nature a hundred years before Bartram. Banister explained naturalists' motivation for rendering nature themselves, rather than relying on trained illustrators. He noted that the species he found in the Virginia colony were "so strange and monstrous" he feared his descriptions would never be believed. "Considering that dryed plants tho illustrated with never so plaine descriptions will but lead the Limner or designer into many errors," Banister informed a colleague, "I betook my self therefore to drawing."[50] Like Bartram's inscription in *Genera Plantarum*, Banister's own mark making served as a visible link between sensory perception and intellectual construct; it authenticated his representations.

To reinforce the absence of a mediating hand in this setting to paper of sensory impressions, Bartram would sometimes append textual descriptions to his graphic work. He did this in his drawing of the balsam pear, as well as in his drawing of rusty staggerbush (*Lyonia ferruginea*), which he referred to as *Andromeda, or Kalmea* (ca. 1773; fig. 4.6). This work might initially appear to align with the standards of Linnaean specimen drawing as established by Georg Ehret. Here Bartram presented a single panicle of the plant, depicted in an indeterminate, shallow space, and focused quite deliberately on its flowers; one can clearly observe the ten stamens and single pistil of each bloom, as befits its Linnaean class and order, *Decandria monogynia*. Yet the flowers are quite unlike the small symmetrical blossoms an observer might expect from plants of these genera, which included various species of laurel and heather in the 1753 edition of *Species Plantarum*.

In Bartram's drawing, the flowers' oversized, gourd-shaped pistils, surrounded by erratically splayed petals, would certainly have seemed unreal. Bartram noted as much, observing in the appended annotation that "These extraordinary appearances of beautifull Flowers are more like fiction or the exertions of an eregular fancy than of Nature." This was not the standard appearance of the plant's flower, he pointed out, and included an example of its bloom at "natural size" for comparison.[51] Located at the upper tip of the branch,

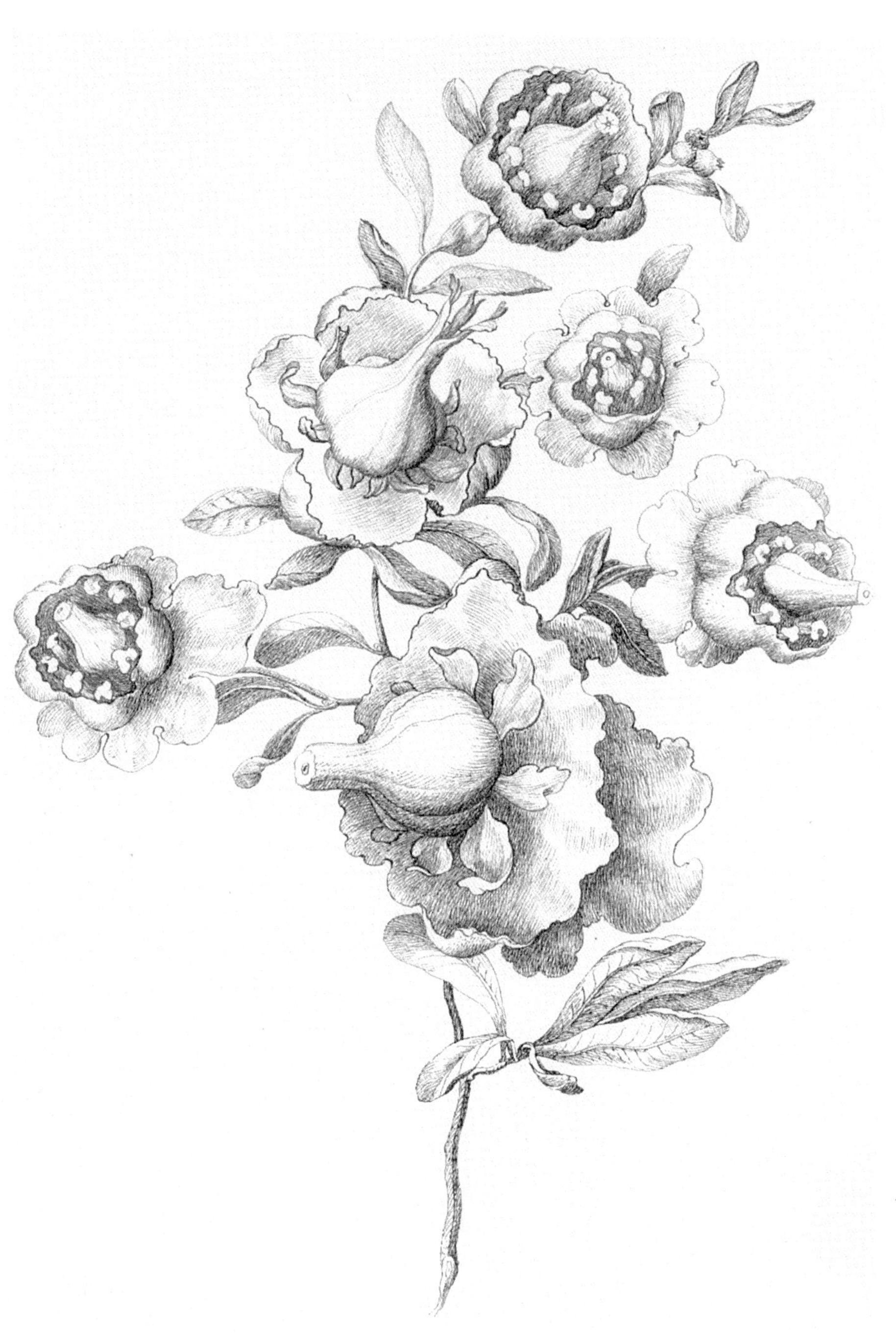

FIG. 4.6.
William Bartram,
*Andromeda, or
Kalmea* [*Lyonia
ferruginea* or rusty
staggerbush],
1773. © Natural
History Museum,
London.

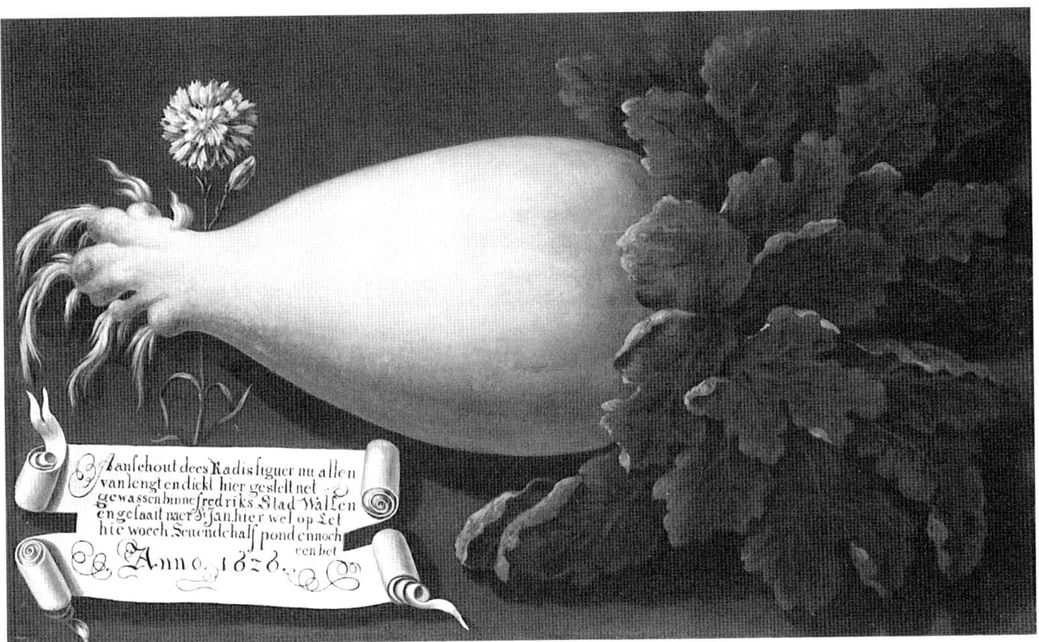

FIG. 4.7. Artist unknown, *A Giant Radish*, 1626, oil on panel. Rijksmuseum, Amsterdam.

the flower looks like a little cup, its corolla pinched tight. Its small symmetrical form highlights the oddity, even indecorousness, of its cousins. Bartram's inclusion of an explanatory gloss, which reinforces the observations as his own, is akin to early modern still-life traditions, as seen in a seventeenth-century portrait of a giant radish. Here, the artist incorporated a trompe l'oeil note recording the place of its growth, its weight, and the year it was found as a way of authenticating the image (fig. 4.7). The notation in *Andromeda, or Kalmea* serves a similar function: even though the drawing portrays a natural object that seems wholly unnatural, the depiction is affirmed by the very person who observed the phenomenon.

Bartram's centering of the plant's flowers in his illustration is typically Linnaean, but his focus on the aberrant blooms is not. Although the Royal Society had followed, in the early part of its history, the Baconian edict to study "monstrosities and prodigious births," by the middle of the eighteenth century the society's aim had pointedly shifted to the study of nature in her regular course.[52] Linnaeus was emblematic of this shift, eschewing atypical specimens in his work to emphasize nature's order, not its exceptions. Linnaeus observed, "The Botanist's science rests on foundations which have been proved and can be proved by certain and infallible geometrical principles. . . . All the species recognized by Botanists came forth from the Almighty Creator's hand, and the number of these is now and always will be exactly the same, while every day new and different florists' species arise from the true species so-called by Botanists."[53] Early in his career, Linnaeus staked out a position that naturally occurring species were static in number, varying only in their nonessential characteristics. Artful crossbreeding of plants by florists and nurserymen could yield so-called new species, but these he assumed would be unable to propagate, and he considered them outside the natural order.

As a result of this perspective, all hybrids and mutations were largely ejected from Linnaeus's taxonomic project. All, that is, save one: the spontaneously arising mutation of the common toadflax (*Linaria vulgaris*) discovered in 1742, which he called *Peloria*, after the Greek for "monster" (now known as *Linaria vulgaris f. peloria*). Common toadflax has four stamens and was situated in the Linnaean class *Didynamia*. The unusual blooms, however, were discovered to have five stamens, a shocking change that demanded a shift of class to *Pentandria*. Nothing, Linnaeus claimed, "can be more fantastic than what has occurred," a natural deviation so profound that "no one can recognize the plant anymore."[54]

Linnaeus authored a dissertation on this supposedly monstrous plant and its potential to upend systems of knowing. In the conclusion, he referred specifically to Trembley's work on the polyp and to corals—which had been recategorized, over time, from stones to vegetables and ultimately to animals—and how they both tested conceptions of nature. Although he admitted that "we still don't understand the cause of such mutation" in the *Peloria*, he hazarded that it was likely the result of cross-fertilization of some kind.[55]

Yet even as a hybrid the *Peloria* was peculiar to Linnaeus since it retained the ability to reproduce, something he assumed impossible. (Linnaeus's *Peloria* results from a genetic mutation, not from cross-fertilization.) Since this monstrosity constituted a new genus and species that could propagate itself, it had the potential to undermine Linnaeus's view of the fixity of the natural order. He called for in-depth study of the plant, since its ability to propagate introduced a bewildering concept—namely, that a single genus could have different organs of fructification. Such a possibility would entirely undercut Linnaeus's taxonomic system and his claim that, inasmuch as his orders and classes might be artificial categories, his genera were entirely natural.[56] After receiving substantial push-back on his theory that *Peloria* might be a spontaneously arising new species, Linnaeus ultimately backpedaled his claims and refused to discuss them further. The encounter with these blooms seems to have shaken him profoundly, and by the end of his career he was no longer doctrinaire about the fixity of species.[57]

Linnaeus's difficulty in accommodating the monstrous toadflax into his system only highlights the ease with which Bartram accepted natural objects that failed to follow existing schema. The staggerbush's curious blooms did not upend his study of nature but drew him deeper into it. He likely assumed the viewers of his drawing would believe the irregular form of the flowers to be a fiction rather than a true representation of nature, so his annotation is intended to authenticate his portrayal and to inoculate it against his audience's doubts. The text indicates that this was something he had witnessed and recorded himself, and indeed, his observations were not mere invention: his drawing portrays flower galls caused by an infecting fungus.[58] Although unaware of what caused the flowers to expand so unexpectedly, Bartram did not attempt to ignore, minimize, or even explain them. Instead, his drawing directs attention to supposed anomalies, but it does so with wonder and aesthetic pleasure. Bartram's attention to this *lusus naturae* or aberration of nature was intended neither to bracket nature in her regular course nor to obsess over irregularities but, rather, to undermine the distinction between them. All of nature's productions were extraordinary and beautiful examples of God's handiwork, and William Bartram fully incorporated supposed aberrations into the compass of nature's design.

The Ornamental Detail

The way Bartram's drawings trigger a living presence response in the viewer, while also authenticating the truth of a particular experience, closely aligns with his view of the natural world as a network of perpetually shifting relationships. In Bartram's telling, organisms are not static or inert but ever responsive to their environment and the contingencies they must negotiate. Although this ecological perspective only became a defined area of study in the nineteenth century, the global exchange and propagation of plants made it an emerging topic in the eighteenth. Even Linnaeus, who devoted his career to isolating and defining differences, recognized the complementary interactions that developed among species.[59] This recognition was admittedly secondary to Linnaean natural history, but it was at the forefront of Bartram's. As Amy Meyers explores in depth, Bartram's natural history pivots from Linnaeus's static, essentializing descriptions of species to an examination of their lived relationships, which he visually conveyed by drawing together environmentally related plants and animals and reinforcing their connection through reflected form.[60] Yet his use of reflected form was only one of his tactics for conveying these ecological relationships. Another and an equally important approach was embedded at the very heart of his graphic practice: his reliance on seemingly inessential or "ornamental" information.

Ornament is often framed as detail that exceeds its descriptive function and becomes aesthetic enhancement, as something that operates along the very edges of representation. For many eighteenth-century naturalists, simplicity and clarity were key to their descriptive enterprises, and ornament itself was often framed as monstrosity. Linnaeus pointedly affirmed this rhetoric of crystalline simplicity, one of nearly complete concordance between object and representation, writing, "Whatever use the flowers of rhetoric may have in language for ornamenting speech at least in the nomenclature of species they are monstrosities: for here we demand the bare and simple truth."[61] As Cynthia Sundberg Wall points out, what constitutes proper description is, of course, historically and culturally contingent. The mere enumeration of objects featured in Daniel Defoe's *Robinson Crusoe* (1719), she observes, would be considered properly descriptive in its historical period, just as the elaborately detailed interiors of Sir Walter Scott's *Ivanhoe* (1820) would be a century later. Wall charts a shift from the narrative terseness of the seventeenth century to the density of the nineteenth and identifies the eighteenth century as a period in which both styles struggled for cultural dominance.[62]

Although Wall focuses on literature, she also addresses other fields of cultural production. She examines the work of Joshua Reynolds, the first president of London's Royal Academy of Art, who promoted a neoclassical Grand Manner of painting that pointedly eschewed antic visual detail. In his *Discourses*, which date to the late 1760s and early 1770s, Reynolds argued that a painting's universalizing principle should never be overwhelmed by particularities, since these reduce the work to "a mere matter of ornament."[63] Reynolds, like Linnaeus, used "ornament" to refer pejoratively to inutile or irregular details that obscure universal truths. Although Reynolds grudgingly allowed that ornament might have its place, he believed too much distracted from, rather than enhanced, representation.[64]

Reynolds's pronouncements were, in part, a response to the highly ornamental Rococo style or "modern taste" that had emerged at mid-century. In both painting and design, the Rococo style was celebrated by some for its inventiveness and sophistication but also derided by others for its incoherence and "lawlessness." The style's French origin made it especially suspicious to British artists and theorists. Even Hogarth, who counted artists and artisans working in this style among his friends and colleagues—and whose own work was very much rife with antic visual detail—loudly disparaged French design as "all gilt and beshit."[65] At the same time, Hogarth's *The Analysis of Beauty* reads as a theoretical scaffolding for this highly ornamental style, one that would justify his use of it while also protecting it from claims of lawlessness. As Anne Puetz explains, Hogarth's *Analysis* is an attempt to articulate a system that would govern the deployment of the Rococo.[66]

Yet even in this supposedly regulatory treatise, Hogarth reveled in ornament. In fact, he attempted to dissolve the inherent opposition between ornament and essence through his use of the serpentine line. As the most expressive and the most beautiful figure, the serpentine line delineates not isolatable essences but ongoing unfolding transformation over time, as revealed in the *Analysis's* explanatory plates. A straight cone twists into a cornucopia; stays and chair legs become increasingly curved and contorted; a pelvic bone transforms into an ornamental figure; and an infant grows up and grows old, while shifting between male and female. The world's mutability is perhaps best captured in "a sort of Lusus naturae" that Hogarth claimed to have discovered growing from an ash tree—a sinuous excrescence whose form conveyed the necessity of a flexuous line over rigidity and rectilinearity.[67] On the Hogarthian model, isolatable, immutable essences do not constitute any universal truth; rather, the only universal truth is the world's perpetual transformation. Bartram's careful attention to the aberrant flowers of the rusty staggerbush—themselves *lusus naturae*—reinforce this perspective.

The tension between static essence and fluid ornament in eighteenth-century visual art finds its natural history counterpart in the difference between Linnaeus and Georges-Louis Leclerc, comte du Buffon, author of the expansive thirty-six-volume *Histoire naturelle* (1749–1804) and a tentative proponent of vital materialist thinking. Where Linnaeus intended to provide universally intelligible descriptions of all known plant species based on a limited set of essential features (a project that resonates in key respects with Reynolds's Grand Manner), Buffon argued that no such essence could be said to exist. His view of nature was relational, centered on an organism's lived behavior. According to Phillip Sloan, the species that Buffon recognized "are not the abstract universals of logic of the taxonomists but are rather systems of concrete relationship between real creatures at the level of physical truth."[68] Buffon's focus on individuality over abstraction is, not surprisingly, analogous to the Rococo or modern taste.

Bartram's graphic work, with its unanticipated flourishes, suggests his incorporation of the Hogarthian Rococo into his visualization of the natural world. Gaudio describes Bartram in a neoclassicizing vein, indicating that he sought out "the most basic forms and patterns that unify nature" and often described objects in simple, geometric terms.[69] In many of his works, circles, ovals, and cones clearly undergird the composition, as if to suggest a more abstract universal truth lies just below the surface. The sparseness of some of his drawings reinforces this point, as though he were omitting unnecessary information in order to present a larger truth with greater clarity. And yet, as much as Bartram's geometrical forms and omitted details might indicate pursuit of an organism's underlying essence, much of the information he presented in his drawings could hardly be considered essential in a Linnaean sense.

Take, for example, his drawing of the balsam pear for Fothergill, which appears at first to hew to Linnaean standards (see fig. 3.7). A comparison with the engraving in *An Illustration of the Sexual System of Linnaeus* (London, 1779), designed by the artist naturalist John Miller, is instructive (fig. 4.8).[70] In his plate Miller wholly dismembered the plant, dissecting and magnifying details that draw special attention to the components of the plant's flower, with eight of the eleven figures highlighting its corolla, calyx, stamens, and pistils. The flower appears to have been flayed with mathematical precision, the better to articulate its key features according to the Linnaean system, while the fruit has been sliced in half to reveal its interior structure. These details seem encased in a hermetic space, interacting neither with one another nor with the viewer. Miller's illustration depicts a balsam pear that has been selected, arranged, and taken apart by the naturalist—a lifeless object to observe rather than a living specimen.

Bartram's drawing demands close, careful attention to locate the flowers. Several are included in the top quarter of the drawing, but they are almost indistinguishable from the leaves; the flowers' calyces are barely discernible and

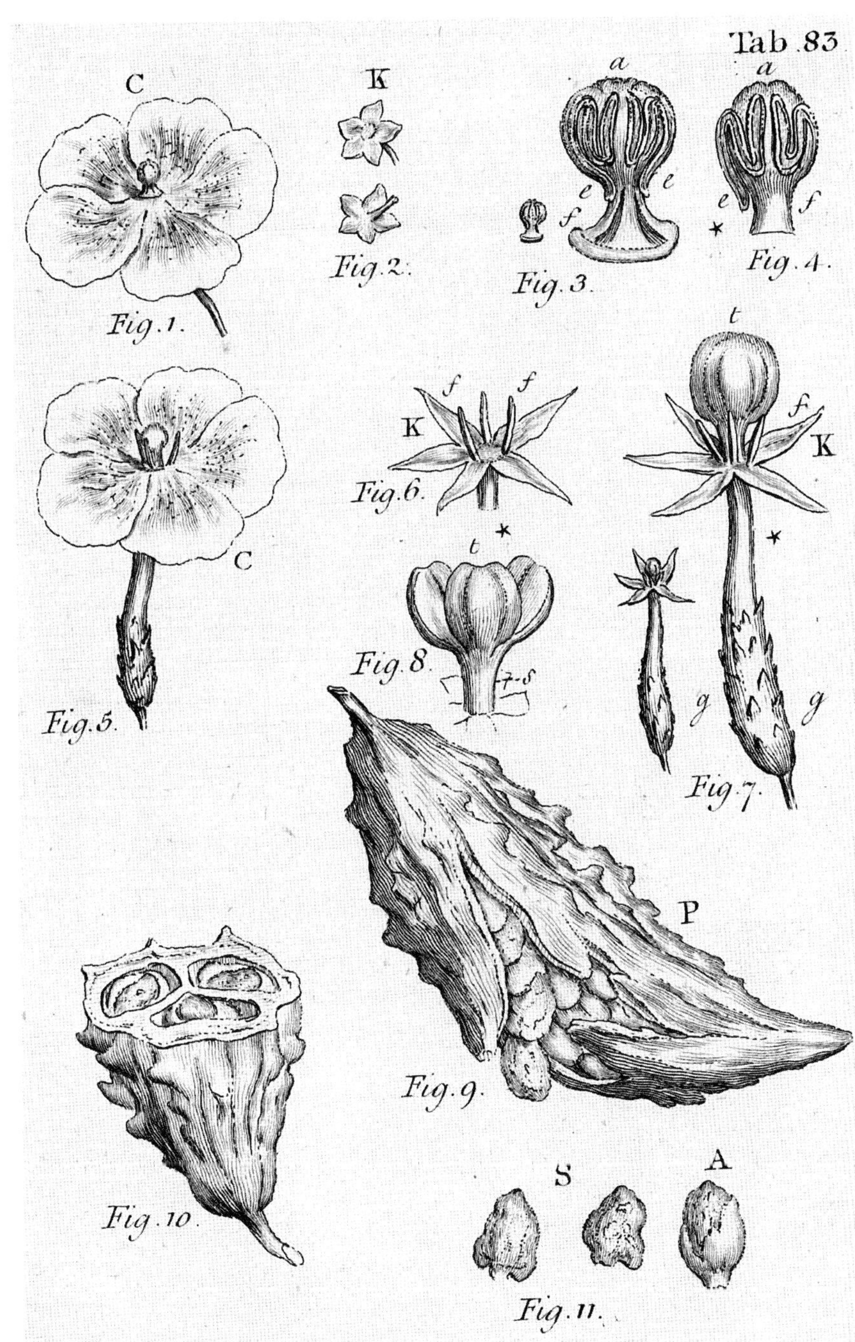

FIG. 4.8. John Miller, *Tab. 83* [*Momordica charantia*]. In *An Illustration of the Sexual System of Linnaeus*, vol. 1 (London, 1779). University of Connecticut Special Collections, Storrs, Connecticut.

their pistils and stamens entirely absent. Instead, Bartram lavished attention on the balsam pear's fruit, handlike leaves, and coiling cirri, which yield little taxonomic information by themselves. Although the details in the composition do not aid in a Linnaean description of the balsam pear, they do convey a sense of how it interacts with its environment. In his *Travels*, Bartram presented the plant as an exemplar of what might be called a proto-ecological view of nature, listing it among those genera whose cirri extend:

> like the fingers of the human hand, reaching to catch hold of what is nearest, just as if they had eyes to see with, and when their hold is fixed, to coil the tendril in a spiral form, by which artifice it becomes more elastic and effectual . . . for every coil adds a portion of strength, and thus collected they are enabled to dilate and contract as occasion or necessity require, and thus by yielding to, and humouring the motion of the limbs and twigs, or other support on which they depend are not so liable to be torn off by sudden blasts of wind.[71]

In this passage Bartram underscored the importance of the lived environment to his natural history and pointed to the role of the ornamental detail—here, the coiling, Hogarthian cirri—in establishing a specimen's interaction with its surroundings. He noted how the waving and spiraling of the *Momordica*'s tendrils add to the plant "a portion of strength," while also allowing it to remain "elastic and effectual," so that it might respond to environmental contingencies. Although he did not portray the specimens with which the balsam pear interacts or the environmental contingencies it must negotiate, he used *S*-curves and spiral flourishes to convey its interactive quality. His balsam pear is not an inert specimen but conveys the plant's lived reality.

That Bartram might have incorporated into his natural history Hogarth's theory of ornament to articulate movement, interaction, and transformation over time is not surprising. Not only was Hogarth's *Analysis* popular in Britain's American colonies, the use of ornament Hogarth promoted would have been evident everywhere in Philadelphia. During the second half of the eighteenth century, the serpentine lines of the modern taste weaved into all aspects of colonial visual culture, from broadsides to sideboards. An especially compelling example is a trade card for the Philadelphia cabinetmaker Benjamin Randolph, which portrays various types of furniture—chairs, bedsteads, writing tables—tucked in among an elaborate asymmetrical cartouche of scrollwork and

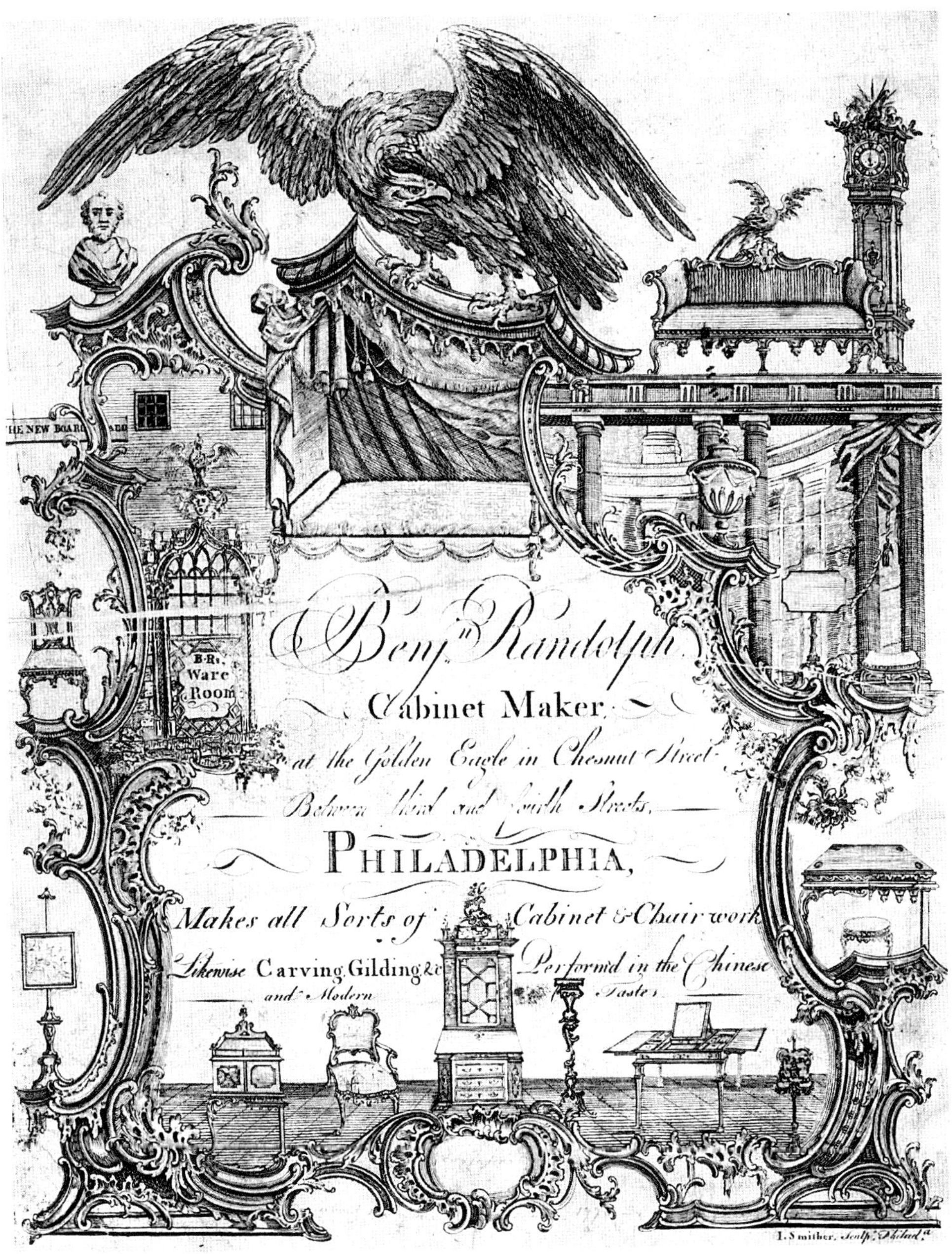

FIG. 4.9. James Smither, *Benjn. Randolph Cabinet Maker, at the Golden Eagle in Chesnut Street between third and fourth Streets, Philadelphia*, 1769. Library Company of Philadelphia.

cascading raffles (1769; fig. 4.9).[72] The extravagant ornamentation is integral to the trade card's meaning: it slows the process of looking, drawing the eye from niche to niche, from side chairs to chests of drawers, as the viewer takes in the full scope of the stylistic variations in Randolph's cabinetmaking enterprise. Just as the ornamental cartouche establishes relationships among the depicted objects and reveals how design elements may be transformed or modified across suites of furniture, Bartram would develop a similar, if more understated, use of ornament in his portrayals of a mutable natural world.

Notably, Bartram's inscription in the copy of *Genera Plantarum*—early though it is—foretells this development. The figure in the center of the flyleaf yokes artist and audience, as the viewer imaginatively traces Bartram's manual gestures. In this, the inscription creates what one might term a representational ecology, but it also represents a mutable natural ecology, of plants and shrubs propagated together. His ornament does not function precisely as the cartouche in Randolph's trade card, which binds together all the productions of his woodworking shop. In Bartram's inscription, this nonmimetic portrayal of animation encourages a process of visual pursuit that opens up a space for imaginative play and interpretive association. The lines pitch and yaw about the page and interact with the claim of ownership above and the words below, which comprise a list of plant genera, taken largely from the Linnaean class and order *Decandria digynia*. Yet, rather than simply re-create the staid lists within the book, the emblem's proximity to these plant names invests them with a sense of motility and underscores their different origins, with species of *Dianthus* and *Hydrangea* from Asia and Europe and *Limodorum* from North America.

The central ornament suggests the wide-ranging social and mercantile network of which the Bartrams were a part; it points to an international web of nurserymen purchasing and shipping, planting and cultivating flowers and shrubs, in order to create the catalogue of genera the emblem adorns. The list begins to take on the appearance of an abstracted eighteenth-century garden plot, in which plants of different origins are propagated together. The imaginative garden Bartram planted on the flyleaf of *Genera Plantarum*, however, was not simply for beauty but also for use and study. It includes *Sinapis*, dowdy mustard plants cultivated for cooking greens, and *Alsine* or henbit, so called for its use as chicken feed.

Bartram's inscription is only suggestive of the relationships among living beings, but the text and figure describe at least one instance of ecological interconnection, at the point where the ornament's lines intersect with the genus *Limodorum*. First identified by Gronovius, the naturalist who gave this copy of *Genera Plantarum* to the Bartram family, the genus had only one species attributed to it: the *Limodorum tuberosum*, now commonly known as the grass pink orchid (*Capologon tuberosus*). Unlike many orchids, it is non-resupinate; its flowers face upward, rather than hanging upside down, which makes them especially alluring to the insects that pollinate them.[73] In Bartram's inscription, the unfurling, almost dissolving form of the ornament pitches about the page and evokes the flitting of insects, an implication heightened by the butterflylike shape that forms where a stippled line alights on *Lymadorum*. The intersection of text and ornament creates a whimsical and unexpectedly mimetic depiction of pollination, offering a delicate ecological vignette that is striking in its scientific and poetic precision.

The conceit of drawing from the life provided Bartram a model for being in and understanding nature. He saw the act of representation not simply as a human activity but as a natural process, by which streams and rills inscribe a landscape and swimming fish create "figures." Of a particular view into the St. Johns River, recorded in his *Travels*, Bartram exclaimed, "What a most beautiful creature is this fish before me! gliding to and fro, and figuring in the still clear waters." The beauty of this particular fish—the great yellow bream (*Chaenobryttus glossus* or warmouth)—depends on the medium in which it swims, which alters the reflection and refraction of its scales. These are "every where variably powdered with red, russet, silver, blue and green specks, so laid on the scales as to appear like real dust or opaque bodies, each apparent particle being so projected by light and shade, and the various attitudes of the fish, as to deceive the sight."[74] In his graphic work, the dynamic relationship between animating mark and receptive ground is not simply a metaphor for his view of nature but, rather, an extension of it.

Bartram's textual description of the great yellow bream—and, especially, his emphasis on visual deception—might suggest that representation can never be a perfect mapping or capturing of the thing as it exists. The medium will invariably alter it, creating a misregistration between image and world. Yet he chose to emphasize rather than obscure these slippages; they are the products of lived experience and underscore that no object or organism possesses a single essential form but is continually shifting, shaped by its environment. In his watercolor drawing of the bream (1774; plate 7), Bartram articulated this idea visually. He traced the contour of the fish in ink, tipping its underbelly forward to offer a fuller view of its body, and deployed a series of small parentheses-like marks to create a scalelike pattern over the bream's surface.[75] At the same time, he incorporated dashes and washes of pigment that only occasionally conform to the inscribed lines that shape and divide the fish's body. With this drawing Bartram suggested order, yet he also upset it. The drawing's patches of color, resolutely unbound by the pattern of lines, convey the deception of the bream's shimmering surface and the variability of sensory experience.

Paradoxically, such disruptions yield a more—not a less—accurate image, pointing to the continuity between the drawing and the world it depicts. The fluid jewellike pigments that form the bream are, like the waters of the St. Johns River, indivisible from it. In his rendering of the fish, drawing is no longer a metaphor, medium, or vehicle for conveying the world but is the world as experienced by the artist naturalist.[76] The image and its creation reveal how artificial the boundary is between the naturalist and the natural world. In Bartram's oeuvre, nature's mysterious, fascinating flux can only be apprehended in and through the world as it is experienced, not through any idealized or abstracted formation of it. The clarity and coherence that Linnaeus established through his taxonomic enterprise ultimately results in the natural world's denervation. In order to capture the liveliness and dynamism of nature, Bartram relied on ornamental figures that were continuous with its persistent mutability.

Chapter 5

Refiguring and Recollecting

Bartram's Legacy

The most productive and original period of William Bartram's career as an artist naturalist was between 1767 and 1777, beginning with his innovative lotus drawing for Peter Collinson and concluding with the end of his travels through the American South. After a botanizing journey that covered thousands of miles and took the greater part of four years, he returned to Pennsylvania in January 1777, in the midst of the Revolutionary War. During his time away, his father had retired and his younger brother John Jr. had taken over the family nursery business, but the start of open hostilities with Britain in 1775 made botanical trade exceptionally difficult. On September 11, 1777, the British defeated General George Washington's troops at Brandywine, and John Sr. feared the conflict would destroy his life's work—his garden. He died on September 22, just four short days before the British occupied Philadelphia. Bartram's Garden was spared during the occupation, but business remained slow until the Treaty of Paris in 1783.[1]

On his return home from his southern sojourn, William worked toward establishing his legacy as a naturalist. He reviewed old notes, travel journals, specimens, and drawings, recombining them in ways that he hoped would introduce his natural history contributions to a broader public. In 1786 he prepared a first draft of the *Travels* for the Philadelphia printer Enoch Story Jr. and had at least nine of his drawings engraved to illustrate it. Bartram and Story were unable to secure enough subscribers, however, and the publication fell through; it was not until 1791 that his *Travels* would come to press with the Philadelphia printing house of James & Johnson. During these years Bartram also prepared a suite of four watercolor drawings for an amateur botanist, the London-based Robert Barclay, likely considering these works for publication as well, either in the newly founded *Botanical Magazine; or, Flower-Garden Display'd* or in the Royal Society's *Philosophical Transactions*. The practice of publishing—of having one's research indelibly printed—in some ways runs counter to a dynamic natural world and the forms of representation it demands. Publishing required Bartram's experiences, observations, and discoveries be winnowed and amended by an outside hand, made less specific and personal and thus less accurate. Yet, in order to have his work broadly recognized, he paradoxically had to relinquish it to others.

Alongside these efforts, Bartram pursued a less institutional method of establishing his legacy, one better suited to him, if perhaps less demonstrably public. For Bartram, acts of mentorship and the shared experience of place served as forms of recollecting that could preserve specific detail and individual experience. His collaboration with Benjamin Smith Barton—who served as the professor of botany at the University of Pennsylvania, after Bartram had declined the position—provided an important opportunity for his work to be refigured and put to use, if made public under another's name.[2] Their joint efforts yielded the publication of the first American botany textbook, *The Elements of Botany, or Outlines of the Natural History of Vegetables* (1803), which contained sections on plant anatomy and physiology and an introduction to Linnaeus's sexual system. Beyond this, Bartram also served as an important interlocutor and guide for a new generation of naturalists wishing to explore the southern landscape. In pursuing both avenues in his efforts to ensure the visibility of his work—his attempts to disseminate his research through publication, as well as his acts of guidance and mentorship—Bartram established a legacy that, much like his natural history, carefully balances the universal and abstract against the local and particular.

A New Set of Figures

The drawings Bartram prepared for Robert Barclay in 1788 can only be considered atypical additions to his oeuvre. These images—which include the four watercolors *Oenothera grandiflora* (large-flower evening primrose), *Hydrangea quercifolia* (oakleaf hydrangea), *Anonymos. Bignonia bracteata* (Georgia feverbark, now called *Pinckneya pubens*), and *Franklinia alatamaha* (Franklin tree)—are curious primarily in how stylistically *uncurious* they are. They do not actively map out the acquisition of knowledge or articulate the flux of a vital, mutable natural world, nor do they necessarily use ornamental lines to engage the viewer's imaginative participation or to construct a network of environmental relations. Instead, their simple and seemingly precise delineation gives them a sense of transparency and self-evidence that many of his other images lack. In portraying the four plants he believed to be his family's most important contributions, Bartram may have hoped that the adoption of a more recognizable visual idiom might aid in the construction of his legacy.

The merchant and brewer Robert Barclay was an important patron of the Bartram family's garden during the post-Revolutionary period. He always referred to America as his "Native Country," collecting American maps, pictures of American subjects, and American botanicals as a sign of his affiliation, though he spent only two years of his adult life there.[3] Born in Philadelphia and named for his great-grandfather, the second colonial governor of East Jersey and a famed Scottish Quaker, at age twelve Robert was sent to London to prepare for a career managing the Barclay family linen business. By the time he reached adulthood, the Barclay family concerns had shifted primarily to banking and brewing, with the purchase of the Anchor Brewery in Southwark in 1781. Robert was a primary partner of this business, renamed Barclay, Perkins & Co., and he helped develop it into one of the most prominent breweries in nineteenth-century London.[4]

Barclay's financial success provided ample resources to pursue his interests in gardening and botany, interests likely inaugurated by his uncle David Barclay. The elder Barclay oversaw his nephew's London education and his entrance into the family business, which connected him to Quakers on both sides of the Atlantic. David was a plant enthusiast, an interest he shared with many of his acquaintances. John Fothergill was his close friend and family physician, and William Logan of Philadelphia was an important correspondent. David managed the European plant orders for Stenton, the Logan family estate outside of Philadelphia, and in turn procured American trees for his own estate at Youngsbury, in Hertfordshire.[5]

Robert Barclay's own passion for plants appears to have been fully developed by 1781, when he moved with his wife and growing family to Clapham Common, outside of London. "Having a garden 70 yards long & 25 wide . . . is very promotive of health," he informed his Philadelphia friend and agent, Thomas Parke, and he observed that one side of the garden "is so well adapted by soil & situation for American Plants that I reserve it for that Purpose."[6] In August 1783, he provided Parke with a list of American plants and queried "is Bartrams not the place to procure them[?]"[7] Parke gently directed his friend toward Humphry Marshall, a close acquaintance of Parke's and a cousin of the Bartrams, who ran his own nursery in Chester County. Barclay was pleased his order received such fastidious attention, but again he mentioned Bartram's Garden: "I hope it may please the Learned to send *Plants* as well as *seeds*, which W[illiam] Logan procured for me from Bartrams when I was at Phila."[8] Barclay had returned to Philadelphia in 1774 to take care of his deceased father's affairs, and he may well have asked Logan to fill a botanical order for David Barclay or another member of his uncle's circle. Whatever the case may be, he had done business with the Bartrams before, and he still saw them as a primary outlet for American plants. Although his extant correspondence does not specify precisely when he began ordering from Bartram's Garden again, it makes clear that he had. In June 1785, he instructed Parke to "tell Bartram & Marshall to send me each a Box of rare plants in the fall as usual."[9] The "Bartram" to whom the letter refers is almost certainly William; even though the garden had descended to John Jr., William was its public face and most of the business correspondence is in his hand.[10]

Bartram may have considered Barclay someone who could fill the void left after John Fothergill's death in 1780, and indeed, Barclay was a logical replacement as a patron. His family was part of Fothergill's London circle and he, like Fothergill, supported projects he thought beneficial to the field of botany. He was a member of numerous societies dedicated to the advancement of natural history, from a perspective both scientific and utilitarian: he was a founding member of the Linnean Society, as well as an active member of London's Horticultural and Zoological Societies. His Bury Hill estate south of London boasted a working farm, allowing him to experiment with agricultural improvements; it included expansive gardens, greenhouses, and hothouses, which, after he relocated there permanently in 1805, featured the enormous variety of plants he had acquired from across the globe.[11]

Barclay counted among his associates the eminent naturalist Sir Joseph Banks, the botanist and gardener William Curtis, and the director of the Royal Botanic Gardens at Kew, William Aiton. Following the death of Carl Linnaeus, Banks—appointed president of the Royal Society in 1778—held significant sway in the identification, naming, and promotion of new plants, and he was often called upon to determine whether genera or species were new. William Curtis was in the midst of publishing his *Flora Londinensis* (1777–1798), a lushly illustrated multivolume compendium of plants native to London's environs, and of spearheading the *Botanical Magazine*, a periodical first issued in February 1787. Likewise, William Aiton was developing a catalogue of the plants at Kew Gardens, the *Hortus Kewensis*, which was intended to track the history of foreign flora in Britain. With publication of Bartram's *Travels* stalled in 1788, Barclay must have seemed ideally positioned to secure alternative opportunities for promoting William's work.

Bartram was surely aware of Barclay's assistance in launching the *Botanical Magazine*, and the 1788 watercolor drawings of the evening primrose, hydrangea, and feverbark closely follow the magazine's visual style. This style aligned with the broader contours of eighteenth-century natural history illustration, but because the periodical was aimed at a general audience, the style was simplified, often omitting the excerpted details of taxonomic characteristics. Bartram's depiction of the *Franklinia* operates in a slightly different vein: at nearly twice the size of the other three drawings and with the plant's parts of fructification highlighted, he likely conceived of this image for a more scientifically minded journal, perhaps the *Philosophical Transactions*, published under the auspices of the Royal Society. Bartram was certain that the tree represented a previously undiscovered genus, and its publication in the *Philosophical Transactions* would have given him priority in the plant's appellation. The incorporation of any of his drawings and descriptions in either publication—each of which had established readerships—would have been valuable. Circulation in the

Botanical Magazine would have ensured the broadest possible reach for his botanical observations, while inclusion in the *Philosophical Transactions* would have bestowed on Bartram the imprimatur of the most learned scientific body in the English-speaking world.

Bartram's shipment of drawings to Barclay, accompanied by thirty-seven specimens, arrived in London some time in February 1789. In a letter to Thomas Parke, dated February 24 of that year, Barclay complained, "I have not yet seen S Bush tho I have been often at the Coffee house to meet him & have not yet rec'd the parcel from him with the Magazines Papers & *color'd drawing* of the Franklinia & dried specimens of plants from W Bartram, which I shall prize very much when I can get them.... [I]n future pray request to have such things deliver'd on arrival."[12] His correspondence indicates that he anxiously awaited the materials from Bartram, who, in turn, was frank about his interests. In a letter accompanying the shipment, he explained: "I collected these specimens amongst many hundred others about 20 years ago ... duplicates of which I sent to Doctor Fothergil; very few of which I find have enterd the Systema Vegetabilium.... I cheerfully offer for the inspection & amusement of the curious, expecting or desiring no other gratuity than the bare mention of my being the discoverer, a reward due for traveling several thousand miles ..., besides suffering sickness cold & hunger."[13]

The *Systema Vegetabilium*, published in regularly updated editions between 1774 and 1830, was a continuation and expansion of the botanical component of Linnaeus's 1735 *Systema Naturae*—the work that established his taxonomic model—and served as an international index of recognized plant names.[14] The first edition of *Systema Vegetabilium* features only plant names determined by Linnaeus, while later editions include those determined after his death.[15] Although Bartram submitted his materials to Barclay "cheerfully," he was clearly concerned that his efforts in the field of botany had so far not been publicly recognized. That his letter also asks for his drawings to be shared with Thomas Walter reinforces the point. Walter's *Flora Caroliniana* had just been published in London, and Barclay sent a copy of the book to Humphry Marshall on October 10, 1788. Bartram seems to have been unaware that Walter, although a Briton, lived in South Carolina. He was, however, aware that the *Flora Caroliniana* featured plants that he had collected on his southern journey but that were now published under another botanist's name. Bartram's correspondence with Barclay underscores a growing urgency to publish his botanical discoveries and introductions, lest they become misidentified or widely circulated with improper attributions.[16]

The *Botanical Magazine* was a modest octavo size and was designed to be a popular work of pleasure, issued monthly at the price of one shilling. The launch of the publication was intended to offset William Curtis's financial losses in the production of his far more costly folio-sized *Flora Londinensis*. As the preface to the first volume of the magazine indicates, it was directed toward the clientele of Curtis's botanic garden at Lambeth-Marsh, which he had established in 1779. Those who paid a guinea per year were allowed to walk in the garden and consult its library, while those paying two guineas were allowed to take roots and seeds of plants propagated there.[17] Because rare plants and exotics were among the garden's most popular, the *Botanical Magazine* focused on ornamentals from abroad rather than utilitarian plants or the local species featured in Curtis's *Flora Londinensis*.[18]

The publication of Bartram's watercolors in the pages of the *Botanical Magazine* would have helped draw greater attention to North American ornamentals that the Bartrams had introduced into British gardens. Although one of the three North American plants featured in the periodical's first volume in 1787—the *Dodecatheon meadia* (sometimes now called *Primula meadia*) or American cowslip (fig. 5.1)—was reintroduced to Britain by John Bartram, he goes unmentioned in the accompanying text. The cowslip was first propagated in the early eighteenth century from seeds sent by John Banister, but these plants eventually died out. John Bartram provided new seeds to Peter Collinson in the late 1730s or early 1740s, and by 1744 the American cowslip was again in bloom in England. In 1746 Collinson encouraged Linnaeus to name the genus *Bartramia*, but he declined, instead choosing the generic name *Dodecatheon* and appending the specific name *meadia* in honor of the physician Richard Mead.[19] The description in the *Botanical Magazine* mentions Banister and Mead but omits John Bartram, an omission his son would have noticed.[20]

William Bartram would have also recognized that his unorthodox, more experimental mode of figuring would not have been accessible to the magazine's broad readership. His visual citations, unhinged scale, inventive use of line, and creation of compositional voids would likely have appeared bewildering and impenetrable to the general reader. In his renderings of the large-flower evening primrose, oakleaf hydrangea, and the Georgia feverbark, he adhered to the

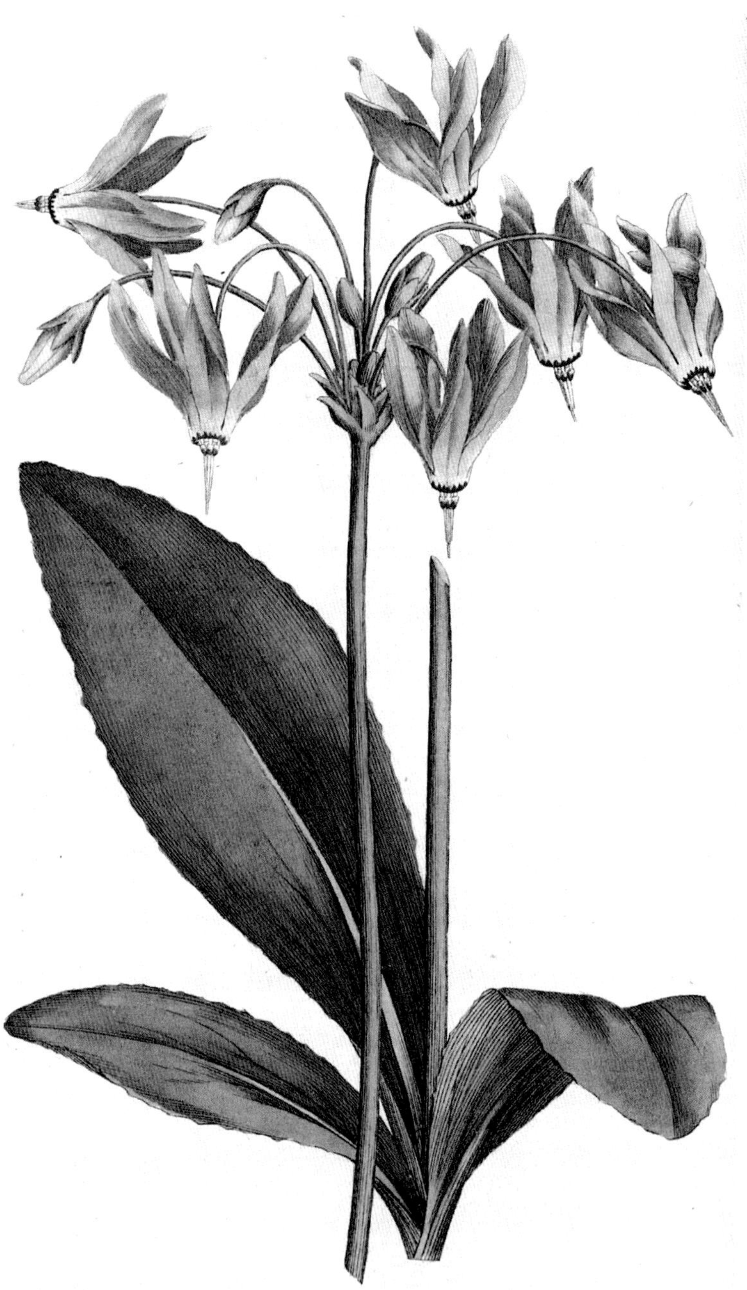

Fig. 5.1. James Sowerby, 12. *Dodecatheon Meadia. Mead's Dodecatheon, or American Cowslip.* From the *Botanical Magazine; or, Flower-Garden Display'd,* vol. 1 (1787). © Florilegius / Bridgeman Images.

magazine's simplified style, which was intended to appeal to the nonspecialist. The *Botanical Magazine*'s plates typically omitted the fussy cross sections or magnified details of a plant's fruit and flower so common to Linnaean illustration and focused instead on its living attitude and color.

Although Linnaeus had found both attitude and color too unstable to be diagnostic, they were key selling points for the *Botanical Magazine*. Curtis noted in the preface that each plate "was drawn always from the living plant, and coloured as near to nature, as the imperfection of colouring will admit."[21] He employed at least thirty artists, engravers, and colorists to ensure the plates' accuracy and consistency. As one reviewer observed, the publication's plates were able to maintain "in perpetual beauty, those fading forms which nature so often produces to be withered and obliterated in the short space of a few hours."[22] The review suggests that illustrations from the *Botanical Magazine* were sought not solely for scientific purposes but also to ornament the interiors of subscribers' homes, much like the living plants that decorated their grounds.

Each plate in the *Botanical Magazine* was accompanied by a textual entry that included the plant's Linnaean class and order, a brief Latin description of its most distinctive features, and bibliographic references to other published accounts of the species. This was followed by a short description in English indicating the plant's origin, how best to propagate it, its time of flowering, and its general appearance. The English-language description helped readers gain a sense of how the plant might be incorporated into their gardens, in accord with the gardening manuals of the day. Bartram's letter to Barclay likewise offers brief taxonomic descriptions of each plant while amplifying their ornamental aspects, aligning with the magazine's focus. He described the primrose, for instance, as a biennial with beautiful yellow blossoms "streaked with flashes or coruscations of flame or vermilion," whereas the hydrangea offered a "gallant appearance even at a considerable distance."[23]

The ornamental thrust of Curtis's *Botanical Magazine* in some respects suited Bartram's aesthetic investment in the natural world. At the same time that he delineated the plants' beauty, however, he also subtly registered his own particular fascination with the inconstancy and mutability of plant species. Even as Bartram classed the primrose as a dazzling ornamental, he also described in detail its unpredictability. He discovered it growing in some old abandoned fields in West Florida, where horticulture had slowly given way to these beautiful, rambling plants.[24] Bartram's narrative suggests a picturesque scene of decay and reclamation, a gloaming set alight with hundreds of golden blossoms. With his watercolor and description of the large-flower evening primrose, Bartram recapitulated themes so common to his natural history: the cycles of life and death, as well as the creation, decimation, and reconstruction of order and form.

So enamored with the flower—which he would describe in *Travels* as "the most pompous and brilliant herbaceous plant yet known to exist"—he collected ripe seeds for Fothergill and himself, planting them in the family garden at Kingsessing.[25] "They grew in equal perfection in the open garden, sowing themselves as an indiginious Plant," he wrote Barclay, "but through inattention we have lost it."[26] His drawing—simplified and standardized though it is—hints at such inconstancy (plate 8). Against the backdrop of the flower's golden petals, the agitated lines of its eight stamens bend and twist away from the pistil and its star-shaped stigma, as though the plant were possessed of two minds. The large-flower evening primrose filters into deserted fields, where it flourishes in the absence of cultivation yet, when propagated, dies of inattention. The proper balance between nature and human intervention can be frustratingly elusive; domesticated nature often behaves quite unlike nature uncultivated.

Bartram's rendering of the oakleaf hydrangea also artfully takes up a theme that courses through his natural history—namely, the potential deception of visual appearances (plate 9). In his letter to Barclay, Bartram described the oakleaf hydrangea as an unusual shrub that possesses leaves like oak trees and flower and fruit like hydrangea. Its proper flowers, however, are not immediately apparent; their tiny yellow and purple corollae are well overshadowed by "large Neutral or Mock flowers" that project from the panicles and can be seen from afar. Closer inspection reveals these damask-hued false flowers have no organs of generation and "seeme designed only to dicorate the plant."[27]

Much of the drawing's composition is taken up by the hydrangea's thinly outlined leaf, with jittery electric lines denoting its "nerves," a representational feature derived from Curtis's publications. In both his *Flora Londinensis* and the *Botanical Magazine*, plates often feature a full stem or panicle of a specimen set before or beside a leaf rendered in outline.[28] Below this, Bartram inserted three tiny ink sketches that capture the true flowers' ten stamens, two pistils, and five petals (though they are unkeyed and go unmentioned in the letter), as well as a delicately tinted watercolor drawing that portrays one of the hydrangea's panicles. Even as his portrayal of the hydrangea depicts leaf, flower, and panicle, this is not necessarily how it reads to someone unfamiliar with the plant. The loose, branching cluster at the lower left does not immediately register as a panicle but, rather, suggests a miniaturized shrub with enormously outsized flowers. It is as though Bartram shrank the hydrangea to convey its full contour yet magnified the false flowers to provide a better view. Even with drawings made to accord with the *Botanical Magazine*'s representational model, Bartram never fully suppressed his philosophical concerns regarding the natural world.

Bartram's visual depiction of the feverbark, like those of the primrose and hydrangea, accentuate the plant's ornamental qualities, yet it differs in some particulars (plate 10). His image and text both carefully portray the shrub's "chiefest gayity"—its snowy white sepals, "stain'd at their tops with an incarnate or Rose blush"—but the drawing appears uncharacteristically flat, particularly in relation to the other two images. Bartram made clear in his letter the reason for this flatness: his drawing was not made from a live plant, as specified for plates in the *Botanical Magazine*, but from a "very perfect dryed Specim[en]."[29] Although Bartram lacked a living sample of the feverbark, he could not forego including it among his observations to Barclay, he explained, because he believed it to be a new genus. To that end, he appended a more scientifically minded textual description to the drawing, which detailed the plant's stamens, pistils, seed, calyx, and corolla.

The inclusion of any of these three watercolors in the *Botanical Magazine* would have been a triumph for Bartram, but the inclusion of the Georgia feverbark especially would have rectified the periodical's omission of his father in its entry on the *Dodecatheon meadia*. The two plants belong to the same Linnaean class and order, *Pentandria monogynia*, and Bartram visually invoked the *Botanical Magazine*'s engraving of the American cowslip in his watercolor. The feverbark's leaves are presented in a similar position to those of the cowslip, and its blossoms are a similar dusty blue. "If it should prove to be a New Genus," he informed Barclay, "I have a Request, that it may be call'd *BARTRAMIA* (*bracteata*) in Memory of My Father John Bartram deceased, The *American Botanist & Naturalist* Whose Laborous Travels . . . hath contributed as much as that of any Man of his time, toward increasing the Stores of Botanical knowledge."[30] More than ten years after his father's death, no botanical production bore the Bartram family name, an oversight William wished to correct.

Naming Names in the Case of the *Franklinia*

As Bartram's plea makes clear, the discovery of new plant species and genera was of great scientific import, but this import was almost always conveyed in affective terms. New genera often bore the name of the botanists who discovered them or of individuals the discovering botanists hoped to honor. Linnaeus adopted this long-held precedent early in his career, and his 1737 publication of the *Hortus Cliffortianus* reads much like a portrait gallery. The plants he selected for illustration are ones typically named for acquaintances, admired historical figures, and sometimes, his enemies. The lemon-scented *Collinsonia*, beautiful blooming *Parkinsonia*, and uncomely *Sigesbeckia* are named for Peter Collinson, Linnaeus's colleague; John Parkinson, author of the *Theatrum botanicum*; and Johann Siegesbeck, a fellow botanist who referred to Linnaeus's sexual system as "loathsome harlotry."[31] Despite Linnaeus's professed desire to remove all metaphors and figurative associations from natural history, his tendency to recognize other botanists in the naming of plants suggests otherwise.[32] And many of his acquaintances saw such recognition as their proper recompense. "Some thing I think was Due to Mee from the Common Wealth of Botany for the great number of plants & Seeds I have annually procur'd from Abroad," Collinson wrote to Linnaeus, "and you have been so good as to pay It, by giving Mee a species of Eternity (Botanically speaking)."[33]

In the letter that accompanied his drawings for Barclay, Bartram observed that he and his father had first discovered the feverbark, along with the Franklin tree, during their first trip to Florida.[34] John Bartram's diary of their 1765 journey notes that they had observed two "curious shrubs" along the banks of the Altamaha River in Georgia but could make no identification since the plants were not in bloom.[35] When William Bartram returned to the area during his own botanizing trip in the 1770s, both the feverbark and the *Franklinia* were laden with flowers, an experience he reported in *Travels*.[36] He collected seed to sow in the family garden back at Kingsessing and prepared specimens to send to Fothergill. Fothergill had made a practice of showing all his specimens to Daniel Solander, Linnaeus's protégé in London, who confirmed both the feverbark and the Franklin tree as new genera, yet little came of the revelation. Solander's interest and attention in the 1770s were largely focused on the botanical specimens from Captain Cook's first South Pacific voyage, and after Fothergill's death in 1780 and Solander's in 1782, Bartram's discoveries lay dormant in London.[37]

Bartram's request that the genus of the feverbark be named for his father is direct but conditional, dependent on the assessment of botanical "grammarians" and systematists.[38] With the Franklin tree, however, he revealed no such doubts regarding its status. Unlike the other drawings for Barclay, which were more suited to the popular and more populist *Botanical Magazine*, he offered a comprehensive visual description of the *Franklinia* that highlighted its taxonomic features according to the Linnaean system. Because his drawing was made from "the living Flowering Tree," which had first blossomed in the Bartram family garden in 1781, he was able to observe the growth of the bloom and the seed vessel's distinctive dehiscence (plate 11). Figure 1 of the drawing depicts the fruit's characteristic mode of splitting, figures 2 and 3 depict the divisions of its interior, and figure 4 portrays the form of its unwinged seeds. To Barclay he wrote "these characters besides many more . . . are I imagine sufficient to destinguish, The *Franklinia* from the *Gordonia* with which I class'd it when first observing it in flower."[39]

The *Gordonia [lasianthus]* to which Bartram referred is the loblolly bay—an evergreen from Carolina with showy white fragrant blossoms, first described in Mark Catesby's *Natural History* as *Alcea Floridiana*. In the 1720s Catesby prepared dried specimens of its leaf and flower for his Oxford patron, but the tree did not gain traction in Britain for another forty years.[40] John Bartram's correspondence records both local and transatlantic interest in the loblolly bay, and his letters often badger his Carolina contacts for seeds and roots. For a time, these contacts included William, during his short-lived tenure as a merchant in North Carolina. John shared the loblolly bay specimens he received from the American South with Collinson, whom he later provided with living plants. Although these plants were damaged while onboard ship, Collinson must have been able to propagate the trees from the remaining roots.[41] When his garden was robbed of his "Most Curious plants" in 1765, he especially regretted the loss of John's "kind present of Loblolly Bays, which throve finely."[42]

The loblolly bay was difficult to transport and to grow, as described in Collinson's letter, which may explain why it was not more popular in British gardens during the first half of the eighteenth century. Only after the London nurseryman John Gordon succeeded in growing the tree from seeds in 1763 did the plant become more widely propagated, as well as more rigorously studied.[43] In his first edition of *Species Plantarum*, Linnaeus, working from dried specimens, identified the tree as a species of *Hypericum*, in the class and order *Polydelphia polyandria*—plants that possess many pistils and stamens, with the stamens clustered into several parcels. When John Ellis observed its living flower in a garden at Clapham, however, he determined it should be reclassified.

In a letter published in the *Philosophical Transactions* in 1770, Ellis placed the loblolly bay in the class and order *Monodelphia polyandria*, or plants with a single pistil and many stamens, evenly distributed. He offered his justification to Linnaeus: "When I compared fresh specimens with the dried ones, it soon became evident . . . why you judged it to be of the class of Polydelphia. . . . For the stamina in the dried specimen appeared to be divided into five distinct phalanges, or bundles, with their filaments united together." He pointed out that the fresh bloom revealed the bundles of stamens to be a result of the drying process, not inherent to the loblolly bay's flower. Resituating the tree in a different Linnaean class and order understandably required the creation of a new genus, for which Ellis suggested the name *Gordonia* "as a compliment to our worthy friend . . . Mr. James Gordon, near Mile-End, to whom the science of botany is highly indebted."[44] Linnaeus, in a return letter, asked that the new genus be called *Lasianthus*, but Ellis demurred: "I am sorry I cannot oblige you in changing the name of the *Gordonia* to *Lasianthus* as it has been presented to the Royal Society," he wrote, indicating that the publication of the new name in the pages of the *Philosophical Transactions* made it definitive.[45] *Gordonia* has remained the accepted genus of the loblolly bay ever since, and Linnaeus dutifully corrected his earlier designation in subsequent editions of *Species Plantarum* and other publications.

Bartram must have anticipated a similar scenario for the *Franklinia*. He and John Jr. first wrote to Linnaeus's son in Uppsala in 1783, with hopes that he might be able to identify the Franklin tree's genus or confirm it as a nondescript and incorporate it into the forthcoming edition of *Systema Vegetabilium*, published in 1784. Bartram explained that he had grown "5 plants, two of which were taken to France by Mons[r]. Gerard Embasedor . . . to be planted in the Royal garden at Versailes. Two plants are here now finely in Flower in the open ground."[46] Although they heard nothing in return (the younger Linnaeus died before their letter arrived), the Bartrams gave a description and specimen to Humphry Marshall as well, who featured it in his 1785 catalogue, *Arbustrum Americanum; or, the American Grove*. Marshall noted Bartram's assessment of it as new: "It seems nearly allied to the Gordonia, to which it has, in some late

Catalogues, been joined: but William Bartram, . . . believing it to be a new Genus, has chosen to honour it with the name of that patron of sciences, and truly great and distinguished character, Dr. Benjamin Franklin."[47]

Marshall was, in fact, dubious that the *Franklinia* represented a new genus (in correspondence, he indicated that he believed it to be a species of *Gordonia*), but several others seemed persuaded of its generic difference.[48] In 1786, when Bartram's young acquaintance Benjamin Smith Barton traveled to Britain to pursue a medical degree, he planned to bring with him "a Box of Seeds of the *most curious* Flowering Shrubs & Trees . . . particularly the Franklinia."[49] While abroad, he also wrote Bartram for a drawing, specimens, and a description of the plant, justifying his request thus: "I mean to visit Upsal, the place where Linnaeus taught . . . and it is customary to publish on a subject relative to *Botany*. I would wish, with your assistance and permission, to publish on the *Franklinia*."[50] Barton regarded William as an important mentor, yet this and other letters suggest that he was often eager to accept honors on his behalf or even to present Bartram's work as his own.

Robert Barclay was similarly enthusiastic about the *Franklinia*. The Bartrams had included it as an "undescript Shrub lately from Florida" in their 1783 plant catalogue, which was published as a broadside in Philadelphia and Paris.[51] Barclay must have ordered it by 1785, since he had successfully grown four *Franklinia* from seeds by the summer of 1787. In a letter from July of that year, he happily informed Bartram that he "carried one of them last week to the Kings Garden at Kew[,] being the first seen there," and in his letter he included an order for more seeds of the Franklin tree.[52] Barclay's visit to the royal gardens and his request for a drawing and description of the *Franklinia* suggest a concerted campaign to introduce a new botanical genus to Britain, an introduction with which Barclay hoped to be associated. With Banks's blessing, the description and drawing of the *Franklinia* might then be published in the pages of the *Philosophical Transactions*, thereby cementing its generic name, as well as the botanical legacies of both William Bartram and Robert Barclay.

It is unclear how long it took Barclay to procure Bartram's drawings and specimens from his erstwhile courier, but he had them in hand by the beginning of May 1789. In a letter dated May 6 of that year, Barclay informed Thomas Parke, that he, Sir Joseph Banks, "& other Gentlemen well vers'd in Botany" had gathered the day before "to examine the Specimens & drawings [Bartram] was so good to send me." Their assessment could only have been disheartening news. "Sr Joseph produced Bartrams works that he had from Dr. Fothergill & on comparing the Franklinia with a former *drawing of Bartrams & dried Specimens* of the *Gordonia pubescens* we were *all* of the opinion that it is the same plant. Still he has WB's observations to reperuse. Pray tell him also that the Bartramia is under consideration, but that he will oblige us by sending us a drawing & specimen of the fruit of that plant."[53]

The *Franklinia* was not considered a new genus, only a synonym for the pubescent loblolly bay, described by Jean-Baptiste Lamarck, Royal Botanist under France's Louis XVI, in volume 2 of his *Encyclopédie méthodique* (1786).[54] The promised reperusal of Bartram's observations of the feverbark—contingent on a rendering of the tree's fruit, which was not included in the drawing for Barclay—reads as an attempt to salve Bartram's disappointment. After receiving Barclay's letter, Parke wrote Humphry Marshall that the *Franklinia* specimens and drawing had arrived safe in London but that "the Botanists in England will not however allow it to be properly named."[55] Sir Joseph Banks conveyed the group's assessment with terse finality in a separate letter to Marshall. The *Franklinia*, he announced, is "a species of *Gordonia*. A drawing of the plant, sent here by Mr. Bartram to Mr. Barclay, has been compared with the specimens; so that no doubt now can remain on that subject."[56]

The *Franklinia* was indeed the same plant as the pubescent loblolly bay, yet it did not necessarily follow that it fell under the genus *Gordonia*, as the European botanists believed. Bartram premised his generic distinction of the *Franklinia* in part on the unusual dehiscence of its fruit, which goes unmentioned by Lamarck and unfigured in the first published illustration of the plant in Antonio José Cavanilles's *Sexta dissertatio botanica* (1788).[57] Illustrations by other late eighteenth- and early nineteenth-century artist naturalists similarly omit this feature. In a plate dated to 1803, the celebrated botanical artist Pierre-Joseph Redouté provided details of the *Gordonia pubescens*'s flower and seed, but none of the splitting fruit, and his 1817–1819 illustration is even less forthcoming, with the fruit portrayed as an inert wooden knob rather than a living organism. Bartram's watercolor for Barclay offers much greater vitality and ultimately more, and more useful, information (plate 11). The leaves of his *Franklinia* wave and reach, bend and dip, while the petals of its corollae curl like tissue. One flower is in perfect bloom; another partially open, its golden stamens revealed; and yet another is a smooth bud. The drawing follows the development of fruit as well as flower, showing the seed vessel in the midst of splitting. Bartram presented not an object but a living process—a process that is essential to determining the plant's true genus.

None of Bartram's watercolors appeared in the pages of the *Botanical Magazine* or the *Philosophical Transactions*. When Barclay sent Bartram a copy of Aiton's *Hortus Kewensis* in 1790, Bartram may have hoped, at least, to find his botanizing efforts recognized there.[58] The first edition of the catalogue consisted of three folio volumes of the plants at Kew, listed according to Linnaean class, order, and genus. The entries include each plant's origin, the date of its introduction to Britain, and the gardener or botanist who introduced it. Although the *Hortus Kewensis* is incorrect on a number of these points, it is an important record of who received recognition from London's botanical community and who went unmentioned. In its first edition, John Bartram is credited with nine plant introductions and William with none. The large-flower evening primrose is listed, but its introduction is attributed to John Fothergill.

In the catalogue's second edition, published between 1811 and 1814, Bartram's Georgia feverbark finally appears, but not as *Bartramia bracteata*, as he had wished. Instead, it is *Pinckneya pubens*, a name coined by André Michaux—like Lamarck, a Royal Botanist to Louis XVI—in honor of the American statesman Charles Cotesworth Pinckney. Its introduction is attributed to John Fraser, the London collaborator for Thomas Walter's *Flora Caroliniana*.[59] Bartram is recognized, however, for the introduction of *Hydrangea quercifolia*, which had finally been featured in the *Botanical Magazine* in 1807, but with an etching designed by the botanical illustrator Sydenham Edwards. This illustration offers none of the whimsical play with scale or rosy tints in the watercolor Bartram made for Barclay, suggesting little of the dazzling visual display that Bartram described.[60]

The *Franklinia* is featured in the *Hortus Kewensis*, but only as *Gordonia pubescens*. The London nurseryman William Malcolm is credited as the source of the plant, though how he acquired it is unknown.[61] Despite attempts by others to affirm Bartram's distinction of the Franklin tree as a new genus, its designation as *Gordonia pubescens* held for more than a century. The botanist Richard Anthony Salisbury, for instance, would declare it a distinct genus in 1803, yet he dismissed the name Bartram gave it. "There is not a shadow of a pretence to call it *Franklinia*," Salisbury claimed, and he christened it *Lacthea* instead.[62] Salisbury's new generic name failed to gain currency, however, and the Franklin tree was almost exclusively classed as *Gordonia pubescens* until the 1930s, when it finally became accepted under the genus Bartram first proposed, *Franklinia*.[63]

Refiguring

As his correspondence with Robert Barclay attests, Bartram was only moderately successful in his attempts to gain institutional recognition through printing and publishing, activities that were ill-suited to him in their impersonality. These forms of public memorialization often originated in close affective bonds and private acts of memory, but they also demanded a degree of forgetting or erasure. The botanical name *Collinsonia* may have been born of the collaboration between Collinson and Linnaeus, but the details of its origin are now only secondary to the plant's generic characteristics. In the move from private to public, specific to general, much is effaced.

Bartram's natural history was dedicated to balancing the particulars of the natural world against a totalizing system of classification and nomination. Typically omitted from such a system are the minutiae of lived experience, but to portray nature as ordered and self-evident fails to represent it accurately. Bartram's drawings often speak directly to the precarious balance between the particular and the general, lived reality and abstract knowledge. Bartram conveyed these concerns both in the subjects he chose to portray and also in the way he chose to portray them. As heavily seamed composites of different representational conventions (the specimen drawing, the picturesque landscape, the map), his drawings highlight the challenges involved in coming to know and portray a dynamic natural world. His emphasis on drawing—on the work of the naturalist's eye, hand, and mind—reveals that the knowledge he presented originated from his particular empirical experience. The sinuous lines and reflected forms in his graphic work force us to trace the progress of his natural knowledge, whereby it becomes our own, just as its voids and blank spots underscore the impossibility of any comprehensive system of knowing.

Notably, most of Bartram's drawings had been intentionally bespoke and semi-private, circulating among an international coterie of fellow naturalists. It was only after the deaths of Peter Collinson, John Bartram, and John Fothergill—when his connection to a transatlantic network of virtuosi became more tenuous—that Bartram pursued a new tack. It was then that he initiated his collaboration with Barclay and finally published his *Travels*, even though this impersonal public legacy ran counter to his natural history. His materials for Barclay quietly register this contrariness. Although they are largely standardized, he appeared unable or unwilling to fully suppress his interest in conveying the richness and complexity of nature. His renderings are lively portraits, not types, and his naming of the Franklin tree underscores this contrarian strain. Bartram presented his European colleagues with a fairly orthodox specimen drawing, but the plant depicted is named for a key instigator of historical and political rupture. Perhaps it is unsurprising then that these materials were not immediately incorporated into institutionally sanctioned natural history in Britain.

The publication of *Travels* was somewhat more successful. Reviewers were drawn to the story but perplexed by its deeply personal style, viewing it more as a *lusus naturae* than a coherent rigorous work of natural history. As Nancy Hoffman indicates, between its first draft and its publication in 1791 the text underwent many changes, in which Bartram attempted to balance personal reflection against public interest. After the 1786 publication fell through, Benjamin Smith Barton offered to publish the work abroad in 1788, but Bartram informed him it was yet "an imperfect Embryo," and claimed to be "doubtful of its consequence in respect to publicke benefit. The Narrative might afford some amusement & serve to Kill time with the inquisitiveness of all denominations."[64] This language, "to kill time," indicates Bartram's concern that the book would fall short of achieving any sort of legacy or self-transcendence. The author Washington Irving would later use a similar construction in describing his own literary production as inconsequential, incapable of securing lasting memory and "the after part of existence."[65]

Memory was no static repository for Bartram, where past experiences might be collated in print or stacked on shelves. John Locke observed that memories "are actually nowhere," except in the mind's capacity "to revive them again; and as it were paint them anew on itself."[66] Only through vigilance and repetition could ideas and experiences be retained, aided by a return to the objects or places that first occasioned them or through continued reflections on those ideas. Building on Locke, William Hogarth conveyed the importance of figuring as a mnemonic device in *The Analysis of Beauty*. Although drawing provided a means of coming to know and understand the world for both artist and viewer alike, it could also serve as an act of remembrance and recollection. "He who will thus take the pains of acquiring perfect ideas . . . of several material points and lines in the surfaces of even the most irregular figures," Hogarth maintained, "will gradually arrive at the knack of recalling them into his mind when the objects themselves are not before him."[67] Hogarth literalized Locke's notion of painting or drawing anew; in the absence of the objects or places that first occasioned ideas and experiences, the act of figuring could reinforce their strength and vividness.

Bartram's mature natural history can be viewed not simply as an act of collection and synthesis but as an act of recollection and resynthesis. His journey for Fothergill in the 1770s was in part a reenactment of the botanizing trip he had taken with his father, nearly ten years before. He revisited people and sites, including the banks of the Altamaha River in Georgia, where he and his father had observed the Franklin tree and feverbark in October 1765. Not only did he collect seeds and specimens of the shrubs on this second trip, he also made figures of each. His watercolor drawings for Barclay refigure these earlier images, adding more detail and color, more substance to shadow. His rendering of the feverbark adds particulars absent from the original drawing made for Fothergill, even though this later work was made from a dried specimen and not in the field, as the earlier version had been. His second version of the *Franklinia* similarly gives the plant more color, weight, and volume.[68] The leaves are tinted deep green, fading to a dusty blue-gray on their downy undersides, and tipped with yellow and orange pigment to convey their deciduous character. Bartram appears to have been attentive to Locke's warning that *"the pictures drawn in our minds, are laid in fading colours; and if not sometimes refreshed, vanish and disappear."*[69] His travels, specimens, and drawings follow closely the Lockean process of memory, of retracing one's steps, reviving and refreshing the ideas and sensations previously impressed on the mind.

Beyond his watercolors for Barclay, Bartram would refigure other objects or sites that he had originally encountered on his travels as part of his ongoing process of remembering. His pen-and-ink rendering of the *Canna indica* (1784; fig. 5.2), or Indian shot, counts among such work. The drawing expertly articulates Bartram's approach to nature: he paid homage to Linnaean taxonomy through the citations in the lower left and right corners, yet he departed from a standard specimen drawing through the addition of an undulating landscape. Rather than show the plant in isolation, Bartram quite literally figured a world around it; and not just any world, but one both disorientingly scaled and delightfully animated. The flower hovers weightlessly in the foreground, and its seed vessel and seeds perch on the grassy bank behind. In comparison to the landscape's tufts of grass and gently nodding trees, the fruit and seeds seem enormous, like Gulliver among the Lilliputians. These ponderous specimens cast shadows on the landscape, but the tiny trees are buoyant and infused with life, bearing remarkable formal similarities to illustrations of freshwater polyps.

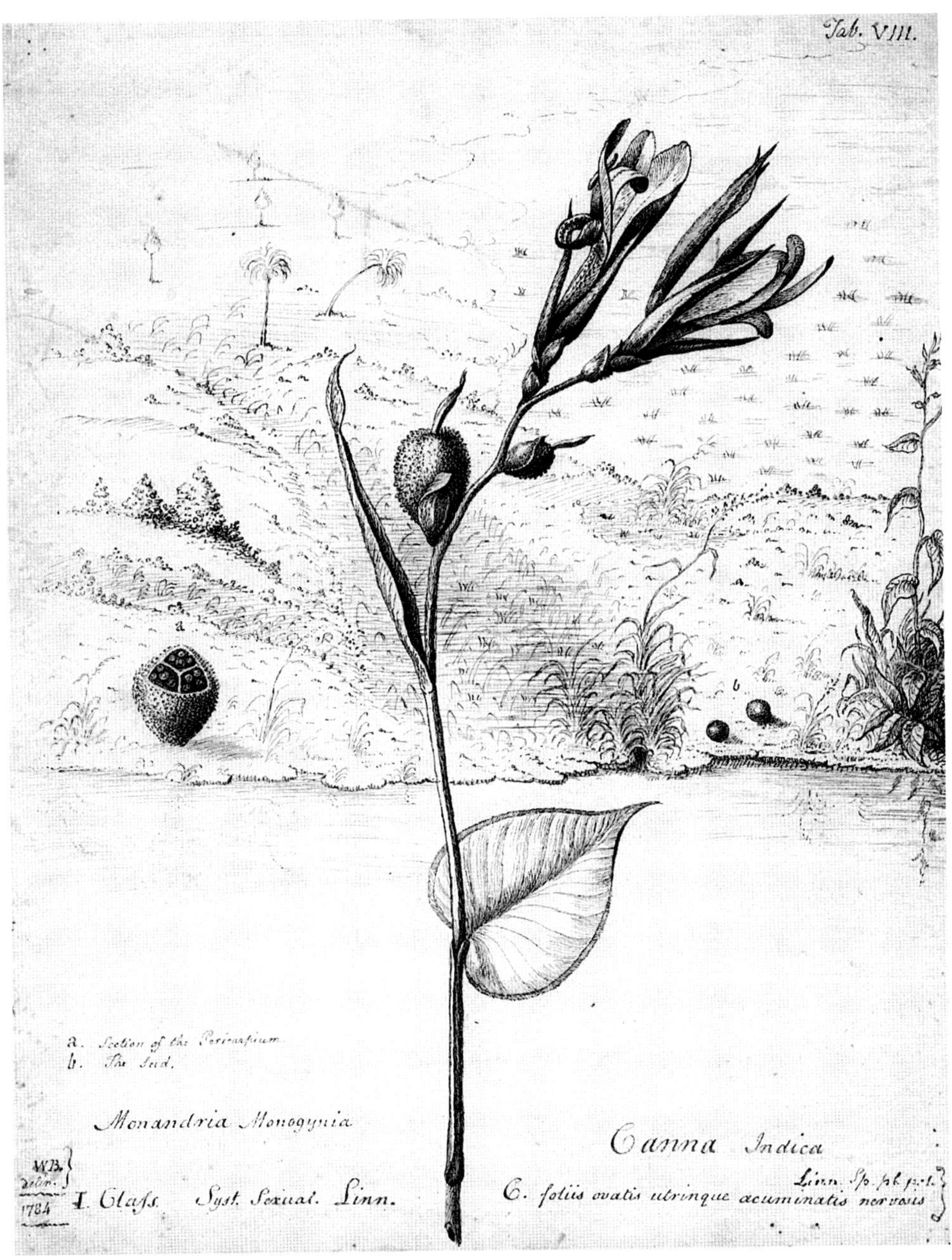

FIG. 5.2. William Bartram, *Canna indica* [Indian shot], 1784. Courtesy of the American Philosophical Society.

In his *Travels*, Bartram described fields of Indian shot along the banks of the Amite River, in what is now Louisiana. The river itself he described as "turgid and stagnate," thick with grass and weeds and exuding noxious vapors. As his party gradually ascended the river, however, the dark funk of the basin gave way to "surprising luxuriance." The Indian shot in particular "present[ed] a glorious shew; the stem rises six, seven and nine feet high, terminating upwards with spikes of scarlet flowers."[70] His drawing of the site—made seven years after his visit—is not precisely mimetic, but it ably captures his memories of the plant's towering quality and its domination of the landscape.

And yet, for these memories to create a legacy, they could not just be refigured by Bartram; they had to filter out to a receptive audience, where others might engage with them. Through the development of a like-minded community of friends and naturalists at the garden in Kingsessing as well as through his expansive epistolary network, Bartram was able to establish such an audience. It was as collaborator and mentor that Bartram's work would be continually revivified and refreshed.

Benjamin Smith Barton played a key role in this process of refiguring. Although contemporaries were suspicious of Barton's eagerness and unconcealed ambition, he also provided Bartram an assurance that his experiences would extend meaningfully beyond his own life. As Amy Meyers points out, William wrote to Barton admiringly, gratefully, for his interest in and pursuit of natural history. He offered "to contribut[e] all I know for its promotion," even while characterizing himself as "comparitively like an old Saw, or Auger; or Ax, worn out, Rusty, & cast away as useless." Barton's drive and ambition made William himself feel "new Steeled, & repared," ready to be put back to use.[71]

Bartram gladly supplied Barton with what observations he had to hand, and there is often a direct line between them and Barton's hastily arranged publications. Barton's *Memoir Concerning the Fascinating Faculty which has been Ascribed to the Rattle-Snake* (1796), *New Views of the Origin of the Tribes of North America* (1797), and *Fragments of the Natural History of Pennsylvania* (1799) all owe Bartram a tremendous debt. Barton's *Fragments* in particular does not draw just on Bartram's observations but also on his approach to natural history. In his introduction, Barton noted that the publication exhibits "a rude and imperfect sketch of the *Natural-History Picture* . . . a picture which, if it were drawn by an able hand, could not fail to prove interesting."[72] Like Bartram, he combined fragments into a sketch that, rude and imperfect as it necessarily must be, would be interesting and worthwhile.[73]

Although Barton's early publications echo Bartram and his approach to natural history, his *Elements of Botany* (1803) is more properly a collaboration between the two. William provided many new drawings to illustrate Barton's textbook, in exchange for Barton's tutelage of his nephew James Howell Bartram, son of John Jr.[74] "Thee obliges us all, in bestowing so much attention & favour to the Young Man, Jamme," William wrote in gratitude. His nephew never completed his medical studies under Barton but instead traveled as a ship's surgeon to Cape Town, Java, and Calcutta in 1804. On his return the following year, he carried with him seeds and plants for his father and uncle, helping to expand the geographical scope of the Bartram family's botanical exchange.[75]

The plates for *Elements* were intended to portray the different classes of the Linnaean system, and Bartram plumbed his memory for illustrative specimens. Many of these drawings refigure plants he had previously observed and described over the course of his long career, particularly those that had special meaning for him. For instance, to illustrate the class *Monandria* (plants with a single stamen), Bartram redrew a blossom of the *Canna flaccida*, or golden canna, which he had first depicted while on his southern travels (fig. 5.3). The drawing he sent to Fothergill incorporates an unusual assortment of natural and manmade goods, including a gastropod shell, an insect, and a carved pipe bowl, this last given him by an aged Creek chief in the Indian village of Mucclasse.[76] The composition is pure Bartram in its pictorial logic, shaped by his interest in conveying the singularity of the region and his experiences there. The depicted objects are not types but, rather, individual portraits whose juxtaposition offers a telling comparison, or perhaps continuity, between the productions of nature and those of culture. As much as he treasured the gift of the pipe bowl, he surely saw the sinuous scalloped edges of the canna's petals, the smooth coil of the gastropod, and the insect's delicately segmented body as equally wonderous, aesthetically pleasing, and affectively resonant.[77]

His new drawing of *Canna flaccida* was ultimately joined by *Hippuris vulgaris*, or common horsetail, and *Canna indica* in plate 8 (fig. 5.4) of *Elements*, illustrating the Linnaean class *Monandria*. The visual portrayal of this botanical class is quite unlike Ehret's illustration of it in the *Tabella* for *Genera Plantarum*, where it is portrayed in wholly abstract terms. Both Bartram and Barton were

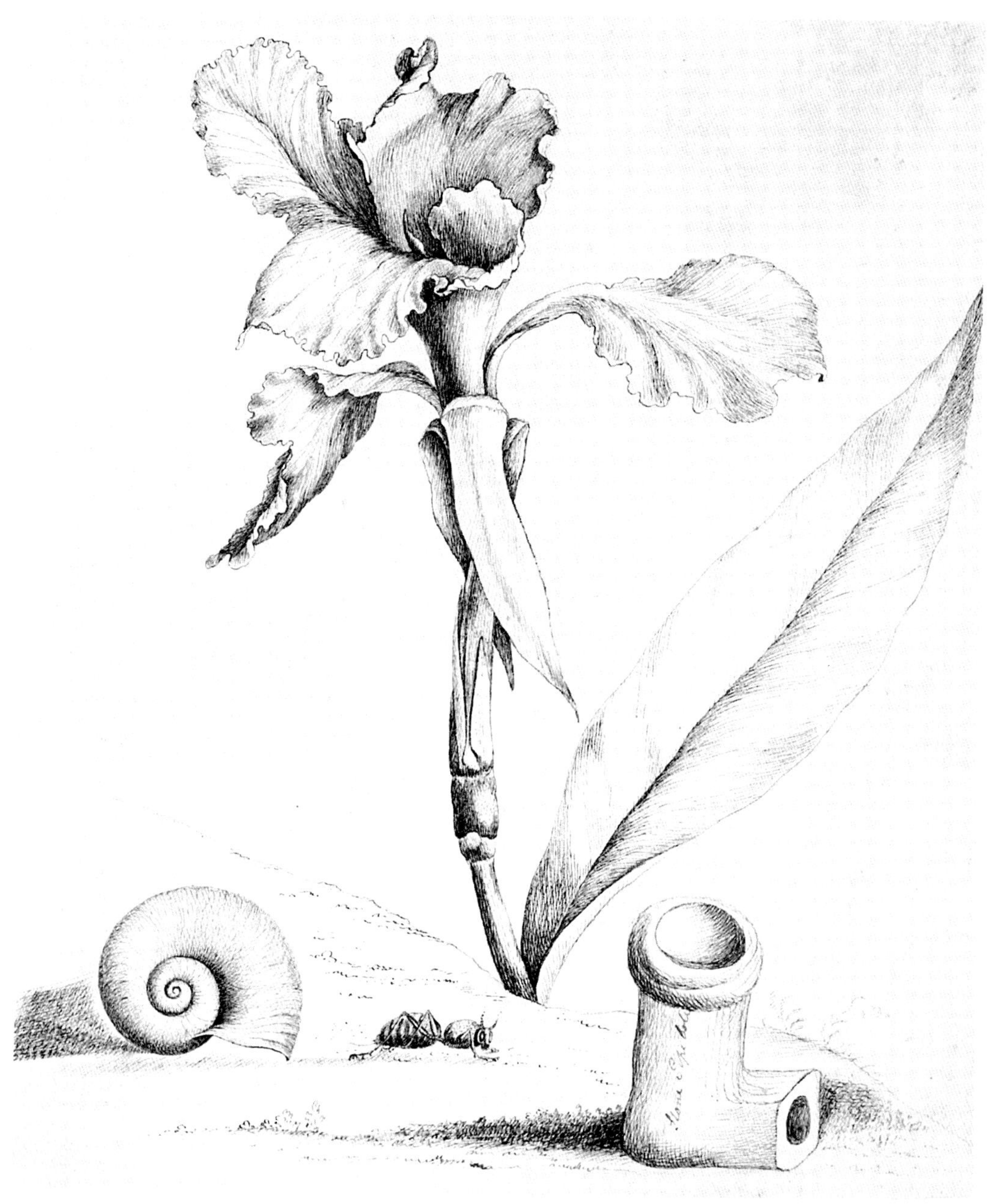

FIG. 5.3. William Bartram, *No. IV* [*Canna flaccida* or golden canna], 1776. © Natural History Museum, London.

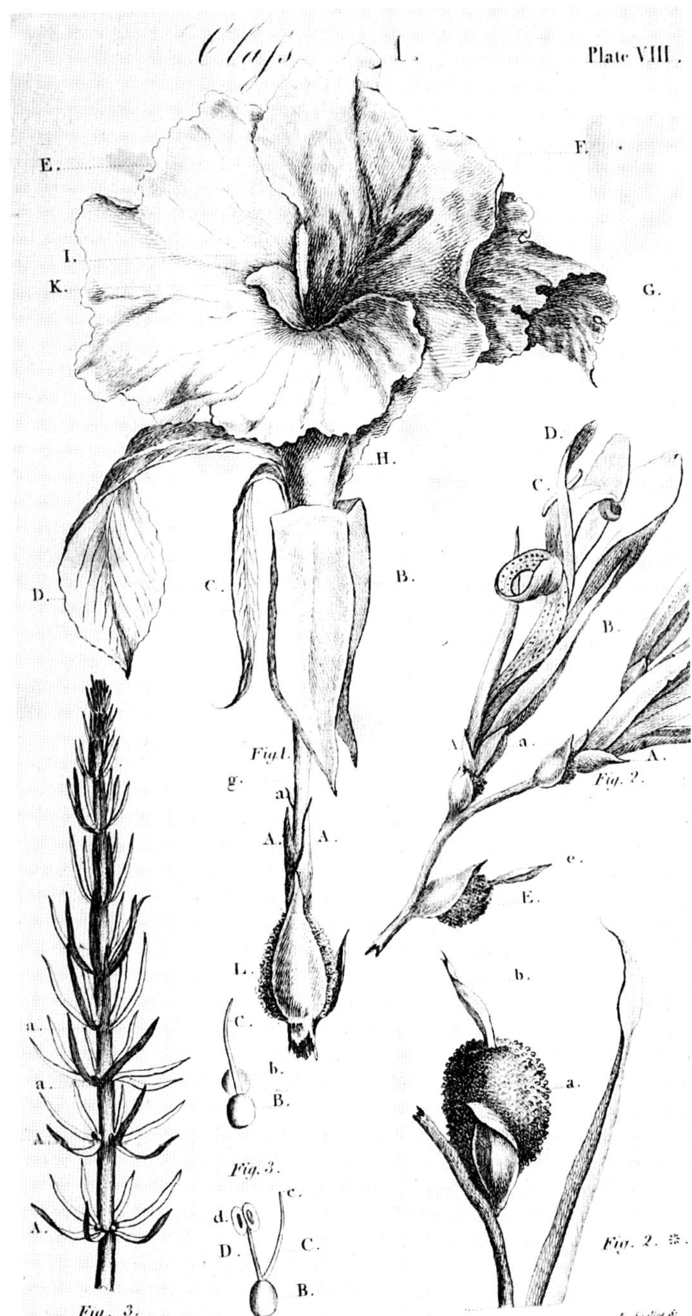

FIG. 5.4. William Bartram (Francis Shallus, engraver), *Plate VIII. Class Monandria.* From Benjamin Smith Barton, *The Elements of Botany*, new ed., revised and expanded by William P. C. Barton (Philadelphia, 1836). US National Library of Medicine.

ambivalent about the artificial nature of Linnaean classes and wished to draw attention to the natural world as a vast array of living beings, not of abstractions. The *Canna indica* featured here is copied from Bartram's 1784 drawing, but its scape is cut in two, the better to fit flower and fruit in the limited space of the printed plate. In pulling apart and recombining Bartram's observations in different configurations, Barton reorganized the natural fragments he gathered from Bartram in ways that suited his purpose yet also paid tribute to their origin. The represented objects maintained their particularity, as they do in Bartram's drawings, but their recombination serves to illustrate taxonomic categories. Although he referred to the publication as "my work," Barton also gave credit to Bartram in the book's preface:

> The greater number of the Plates, by which the work is illustrated, have been engraved from the original drawings of Mr. William Bartram. . . . While I thus publically return my thanks to this ingenious naturalist for his kind liberality in enriching my work, I sincerely rejoice to have an opportunity of declaring how much of my happiness in the study of natural history has been owing to my acquaintance with him; how often I have availed myself of his knowledge in the investigation of the natural productions of our native country.[78]

Re-collecting

While he was working on *The Elements of Botany*, William Bartram began to experience troubles with his vision, which he traced to a fever suffered during his southern travels.[79] In a letter dated October 24, 1801, Barton sent him a letter at Kingsessing explaining his desire for an illustrative example of the Linnaean class and order *Syngenesia monogamia*—which include species of lobelia, violet, and pansies—to feature in *Elements*. Hoping that William might provide a new drawing, he asked, "Do we not sometimes find, at this season, some of the spring Violets in bloom?" William responded the following day, yet the letter is written in his niece's hand, not his own. "I received your note . . . but was not able to read it," he explained, "therefore got my niece to read it to me . . . my Dear friend the exceeding painfulness and weakness in my eyes prevents me from that pleasure and indeed I am fearfull I shall no more be able to make poor drawings for your amusement."[80] His collaboration with Barton on *The Elements of Botany* would mark the end of his career as an artist.

Bartram's diminished vision was part of the reason he turned down other requests and invitations, such as the opportunity to participate in the federal expedition of Red River in 1806. President Thomas Jefferson wished to assess territories acquired through the 1803 Louisiana Purchase, and William Bartram's name was floated for the role of expedition botanist. Barton wrote William a cajoling letter, inviting him to consider the proposition seriously: the journey would be by water rather than foot, the compensation would be significant, and the terms of participation easy. "Come on," he wrote. "You are not too old. . . . You will render great and new services to *natural Science.*" But Bartram demurred. To Jefferson he offered his deep gratitude but explained that he would be unable to participate "on account of my advanced Age, & consequent infirmities. . . . And my Eyesight declining Dayly."[81] He had suffered a severe leg fracture in 1786 that had healed imperfectly, which, together with his failing vision, conspired to keep him in his garden.

Despite this, Bartram remained actively committed to studying the natural world. Unable to draw and no longer traveling, he continued to serve as an important correspondent and interlocutor for other naturalists. Many of them gathered regularly at Bartram's Garden, which would come to be known as the "Botanical Academy of Pennsylvania."[82] The local clergyman and botanist Henry Muhlenberg engaged Bartram in in-depth correspondence about plants, and Barton regularly used Bartram's Garden as a classroom for his students. It also drew naturalists from abroad, such as André Michaux—the botanist attributed with discovering the Georgia feverbark—and his son, François-André. Even nonspecialists were eager to visit, hoping to see the grounds and to meet Bartram himself, the "Philosopher of Kingsessing." Guests to Bartram's Garden ranged from politicians to playwrights: George Washington, the Portuguese ambassador Hipolito José da Costa, and the author Charles Brockden Brown were just a few of the noted personages who spent time there.[83] Through the development of this community Bartram re-created in some measure Peter Collinson's network of virtuosi, but on the American side of the Atlantic.

Different visitors had markedly different assessments of the garden and its various plots. It was organized into ecological niches, ranging from the swampy area near the river's edge (good for cypress trees) to the pond and the unmanicured beds that populated the middle part of the garden. Unlike most water features in eighteenth-century gardens, the pond was murky, intended as a habitat for water plants and amphibians rather than as a visual statement. In the upper garden, beds for flowers, herbs, and vegetables were laid out in a more regular fashion. Some visitors, who anticipated that the garden's design would draw on the manicured style found on English estates (a style adopted, for example, by William Hamilton at The Woodlands) were nonplussed with its appearance. George Washington, for instance, commended the "many curious pl[an]ts Shrubs & trees, many of which are exotics," but he believed the garden "was not laid off with much taste." For a younger generation of naturalists, however, it was a rich laboratory of flora and fauna, stewarded by a generous and unassuming William Bartram, happy to share the details of his life's work.[84]

Bartram's mentees were eager to visit not only the garden but also the sites where he had botanized during his travels through the American South. The refiguring that he engaged in by putting his drawings and observations in Barton's service thus took on a more performative aspect, as younger naturalists and nurserymen traced his footsteps.[85] Some of them hoped simply to participate in the recollection of Bartram's southern journey—refreshing, in Lockean terms, the pictures in the mind by tracking down key locales and observing the flora and fauna in situ. These naturalists often brought their observations to share with Bartram specifically or published them in one of the emergent scientific journals of the day. For instance, when William Baldwin, a Philadelphia medical student, moved to Georgia in 1811, Henry Muhlenberg reminded him that he was in "an excellent situation to elucidate Bartram's TRAVELS" and encouraged him to write up his observations on any "dubious plants."[86] Was the flora Bartram described real or was it mere fancy? Some plants were difficult for Baldwin to trace, but many others he was able to confirm. When he paid a visit to Kingsessing in 1817 to share his observations, Bartram felt happily vindicated and deeply grateful.[87]

Others who followed in Bartram's footsteps were more commercially minded, looking to collect specimens for shipment or sale. Such was the case with André Michaux, who had established a garden in New Jersey to collect and propagate plants that could then be sent to France. He was not content, however, to trade in species that were already well known; instead, he wished to play a role in the introduction of the new and the novel. Understanding that the American South might boast flora better suited to the French climate, he determined to establish a second garden near Charleston. Before heading to South Carolina, he visited Bartram's Garden, where he made note of the southern plants propagated there, particularly the *Franklinia,* and learned details of William's own botanizing travels a decade earlier. Michaux spent much of the next ten years in the South, developing his garden, searching for plants, and sending seeds back to botanists in France and the northern United States.

In the course of his botanizing, Michaux revisited locales previously explored by Bartram, and his journals make mention of plants that William had either shown him or described to him. On May 2, 1787, he described the "*Nyssa ogechee* [Ogeechee tupelo] of Bartram," observed along Georgia's Ogechee River. On December 6, 1788, while traveling along the Keowee River in South Carolina, he took note of the mountain magnolia, which Bartram had called *Magnolia auriculata.*[88] Bartram had an engraving made of the plant for the 1786 edition of *Travels*, which he very likely shared with Michaux during his visit that June. *Travels* was not published until 1791, so Bartram was not recorded as the discoverer of this plant either; again, that recognition went to Thomas Walter, who had published its description in *Flora Caroliniana* and named it *Magnolia fraseri.*[89] Although Michaux was successful in locating the Ogeechee tupelo and the mountain magnolia, he seems to have had little luck in tracking down the coveted *Franklinia.*[90]

Even though there was debate about the generic status of the tree, it was still a beautiful ornamental, eagerly sought by plant enthusiasts both in the United States and abroad. The English nursery of Grimwood, Hudson & Barritt, responding to this interest, had urgently requested that Humphry Marshall send them "as many as you can of Franklinia."[91] Since Bartram was the only source of the *Franklinia* outside of the wild, the following year Marshall sent a nurseryman to Georgia to search for the tree, only to learn via letter that "there is not a plant of the Franklinia to be found the French Kings Botanist [Michaux] have been several times looking for it where it is said to grow also a Gentelman from London but found non it growes as reported 200 miles to the south of this."[92] Marshall's nurseryman may have been off course or unable to recognize the plant in its natural habitat. Marshall's nephew Moses was able to locate the *Franklinia* in 1790 while on a botanizing trip for Sir Joseph Banks; he observed it growing on the banks of the Altamaha River, but he noted there were challenges in transporting it:

When in Georgia I procured & brought home about 20 Plants of Franklinia but 5 or 6 of which are now living & those so weak I have thought best not to forward them ~~this time~~ at present. Wm Bartram the only possessor of it here, or at least who is yet able to propagate it, will not part with any except for Europe & then with only one or two Plants with each Collection—In order to serve you I . . . have procured from him a small Box with two Franklinias & a few other Plants perhaps not unacceptable.[93]

John Lyon, a gardener for The Woodlands estate outside of Philadelphia, would similarly retrace Bartram's steps in 1803, this time using a copy of his *Travels* as a field guide. While trekking through Georgia and Florida, he maintained a journal in which he noted points of alignment between his travels and Bartram's *Travels*, taking particular care to seek out plants Bartram had so enthusiastically described. He found the rusty staggerbush, the oakleaf hydrangea, the Georgia feverbark, as well as the *Franklinia*. "In the bottom between two sand hills . . . grows Bartram's Franklinia," Lyon reported. "It has never been found growing naturally in any other part of the United States as far as I can learn, and here there is not more than 6 or 8 full-grown trees."[94] Yet with each subsequent botanizing trip to Georgia, the number of *Franklinia* seemed to dwindle: Bartram encountered two or three acres in 1777, and twenty-five years later, John Lyon noted only a handful. The *Franklinia* now exists only in cultivation, direct descendants of the seeds Bartram propagated at Kingsessing.

Although frustrated during his lifetime in his attempts to establish his legacy through institutional recognition and the denomination of new genera, Bartram's acts of mentorship generated an alternative path. Other naturalists were eager to seek out the plants he had introduced to them at Bartram's Garden, and many followed his steps through the American South, re-creating his natural history excursions. After his death, he achieved the institutional recognition he sought, though it was by a circuitous and meandering route. Perhaps most notable is Bartram's inextricable link to the Franklin tree, which now not only bears the taxonomic designation he developed for it but is indexically, genealogically associated with him. As the plant dwindled in the wild, Bartram came to replace the banks of the Altamaha River as the *Franklinia*'s origin. Through his curious drawings and his devoted mentorship of a younger generation of naturalists, Bartram ultimately figured for himself an idiosyncratic place in the annals of natural history, one that balanced the universal and abstract totalizing system he hoped to inform against his own very local and particular experience of the natural world.

Conclusion

Nature Animated

Much of William Bartram's legacy ultimately derived from the publication of his *Travels* and his acts of mentorship. Why, then, consider his graphic work at such length? The reach of most of his drawings—such as those made for Peter Collinson, John Fothergill, Robert Barclay, and Benjamin Smith Barton—was constrained in his lifetime, primarily limited to his patrons and their inner circles.

When Collinson died in 1768, his cache of Bartram materials descended to his son and grandson before being sold at auction in 1834. The drawings were purchased first by the antiquarian Aylmer Bourke Lambert and later by Edward Smith Stanley, the thirteenth earl of Derby. Lord Derby incorporated them into his private collection of natural history materials at Knowsley Hall, his estate outside Liverpool, and the drawings descended within the family until their acquisition by an American gallerist in 2021. Fothergill's collection was auctioned off upon his death in 1780, when most of his Bartram drawings were acquired by Sir Joseph Banks. In 1827 these were given to the Natural History division of the British Museum (now the Natural History Museum, London), in accordance with Banks's will. The watercolors that William sent to Robert Barclay made their way there, too, but not until 1914.[1]

Bartram's drawings for Benjamin Smith Barton also remained in private hands well into the twentieth century; they were deposited at the American Philosophical Society only in 1970. As a result, Bartram's graphic work—admired by his patrons for both its content and its curious beauty—did little to shape the field of botanical or zoological illustration. His works are not typically included in the general histories; when they are, they have been considered charming, if amateurish, outliers. Drawings so animated and animating had been rendered inert both by their inaccessibility and by their very singularity.

It was not until the publication of the 1968 facsimile edition of the Bartram materials at London's Natural History Museum that his graphic work began to receive more serious scholarly investigation. His drawings have since been studied as important documents in the history of science, either for offering the first representation of a species or for articulating an emergent ecological worldview. Bartram's 1767 depiction of the American lotus is an example of both. Yet Bartram's graphic enterprise did more than generate documents in the history of botany and zoology; it also modeled his understanding of a dynamic natural world. His images are important not just in what they represent but in how they represent it.

The pleasure Bartram gained in exploring the natural world derived from nature's tendency to confound, exceed, or evade expectations. In a proto-ecological understanding rooted in his empirical experience of nature, Bartram saw it as a series of interconnected, shifting relationships rather than as a static repository of isolated and immutable essences. His natural history research made clear that the biota he encountered were not separate from one another but were engaged in perpetual collaborations, competitions, and negotiations. Bartram's botanizing travels made clear that he, too, was not distinct from these processes, and it was to the natural world's various involutions and permutations—its operations as much as its objects—that he turned his attention. In his most compelling drawings he similarly refused to offer what an audience might anticipate; viewers must actively work to make sense of his often unexpected aggregation of depicted objects, his play with scale, and his ornamental flourishes. Yet Bartram's graphic work is also so carefully composed that such evasion can hardly be accidental. Instead, the drawings send viewers off on their own explorations, mimicking how Bartram came to know and understand such a dynamic natural world.

Throughout this book, I have described the representational conventions of seventeenth- and eighteenth-century natural history—whether the composite views of Robert Hooke, the reasoned images of Georg Ehret, or the naturalizing tableaux of George Edwards—as deriving from the guiding principles of transparency and self-evidence, as though the natural world simply revealed itself to the observer. The aim of such images was to make objects legible, to standardize them visually, to present nature as governed by rules rather than by exceptions.[2] Bartram eschewed such a model of transparency, which minimized the interrelationships he observed among living organisms and effaced the efforts he made to collect and study them.

Against such standards, his drawings have been assessed and often found wanting. Yet standardization was not Bartram's primary interest, nor was his end goal the creation of images that would be easily legible. It was the vitality of the natural world that most captivated him. What happens when the living form one observes yields no immutable essence—or when that essence is in fact its very mutability? How might naturalists convey the entwined relationships that structure the natural world—or their envelopment within those structures? Representational transparency in this instance would take on a dramatically different expression, one in which nature is not shown to be self-evident but, rather, a beautiful and bewildering confluence of forces. Rather than appear as windows onto the natural world, such portrayals would need to entice and confound, eliciting the viewer's active intellectual and imaginative engagement so as to generate a sense of animation and liveliness. To craft such vibrant and vigorous images of nature required the conjoining of different visual conventions, a focused pursuit of optical inconsistency rather than standardization. This, I argue, is the approach Bartram took in his natural history drawings, the deployment of an alternative mode of transparency that conveys the lived messiness of a dynamic cosmos.

Although Bartram's approach was inventive, he had models from science and art to draw on and to adapt to his purposes. Published depictions of the freshwater polyp for instance, generated by Abraham Trembley's research in the 1740s, would have provided a useful template. Since the polyp had no stable form, it required the illustrator to present a multiplicity of views, conjoining various scales and perspectives to capture the polyp's inherent variability and uncertain conceptual status. Perhaps more foundational, however, were the Lines of Beauty and Grace in William Hogarth's *The Analysis of Beauty*—their delineation of ongoing, unfolding transformation over time, as well as their ability to lead the viewer in optical and intellectual pursuit. In locating the origin of the ornamental flourish in the *lusus naturae* or natural aberration, Hogarth indicated that he saw the world as composed of continual change, not of immutable forms. Moreover, Hogarth's text makes clear that observers are not—cannot be—either distanced from or disinterested in such transformations, since they are by necessity enmeshed in the very world they observe.

Ostensibly an aesthetic treatise, Hogarth's *Analysis* is as much about recording nature as John Woodward's *Brief Instructions for Making Observations in All Parts of the World*, the text Peter Collinson recommended to John Bartram in their early correspondence. That Hogarth himself would have considered the *Analysis* in this light is not so far-fetched—he was, along with Collinson, a governor of London's Foundling Hospital, and he also worked closely with Fellows of the Royal Society. Hogarth counted among his friends and correspondents John Ellis, the naturalist who published the first full taxonomic description of the Venus flytrap and established the genus *Gordonia*. Indeed, Hogarth signaled a desire to shape the standards of scientific imaging by depositing a copy of his *Analysis* with the Royal Society's librarian upon its publication. Although his potential influence on natural history illustration has been little studied, there are tantalizing links. Sydney Parkinson, the artist who traveled on Captain Cook's first Pacific voyage, is known to have brought a copy of *The Analysis of Beauty* with him, and the physician and naturalist Erasmus Darwin made explicit reference to the Line of Beauty in his *Zoonomia; or, The Laws of Organic Life* (1794–1796).[3]

As I propose throughout this book, Hogarth's treatise likely served as an important source for William Bartram as well, providing him with a model for coming to know the natural world and a means for portraying its mutability and irrepressible vitality. Still, the era's broader scientific infrastructure demanded images of natural objects that were easily legible rather than examples of Hogarthian discovery and perpetual flux. Relatively few of Bartram's drawings were engraved or otherwise made public during his lifetime; those that were made public were typically excerpted, re-crafted, or otherwise restrained to fit the aims of descriptive botany and zoology.

This is not to say that no other naturalists of the eighteenth and early nineteenth centuries investigated the issues that animated Bartram. He was very much a part of an emerging wave in the natural sciences that explored questions of agency, mutability, and site specificity. These were central concerns for his contemporaries Erasmus Darwin, Jean-Baptiste Lamarck, and Alexander von Humboldt, yet these concerns were seldom explored through drawing or made manifest in pictorial form. Almost none of his contemporaries in their published works deployed a visual language that, like Bartram's, was so earnestly devoted to conveying the dynamism of nature. Although it is beyond the scope of this book to offer in-depth analyses of these naturalists, it is worth briefly considering how they addressed questions of a dynamic natural world in their work.

Darwin, based in the English Midlands, developed his interest in botany relatively late—about the 1770s—when he began translating Carl Linnaeus's writings first into English and then into verse. As his poetry makes clear, his focus was less on Linnaeus's essentialism than on the poetic possibilities inherent in his sexual system. In 1789 Darwin anonymously published *The Loves of the Plants* (reprinted as part of his *Botanic Garden*, in 1791, under his own name), a poem in heroic couplets that examined the romantic assignations of different flora. As it was for Bartram, Ovid's *Metamorphoses* was a touchstone for Darwin, who wrote that the purpose of his work was a patently Ovidian one. In his preface, he observed that just as Ovid "did by art poetic transmute Men, Women, and even Gods and Goddesses, into Trees and Flowers; I have undertaken by similar art to restore some of them to their original animality." Throughout the poem that follows, Darwin invested various plant species with agency, as they preen, flirt, and seduce one another. Of the American cowslip (*Dodecatheon meadia*), with its five stamens to a single pistil, Darwin wrote: "MEADIA's soft chains *five* suppliant beaux confess, / And hand in hand the laughing belle address; / Alike to all, she bows with wanton air, / Rolls her dark eye, and waves her golden hair." Curiously, these couplets are footnoted to reinforce their scientific, rather than merely literary, value. In the note appended to the cowslip stanza, Darwin remarked that "as soon as the seeds are formed, it erects all the flowers-stalks to prevent them from sailing out. . . . Is this a mechanical effect, or does it indicate a vegetable storge [filial protectiveness] to preserve its offspring?"[4]

Darwin and Bartram clearly shared an interest in plant motion and volition, and what it might suggest about the order and structure of organic life. Still, the plates used to illustrate *The Loves of the Plants* are very much the images one might encounter in any botanical publication of the period. Darwin's illustrator,

Frederick Nodder, was also the artist for *The Naturalist's Miscellany* (a journal akin to the *Botanical Magazine* but broadened to encompass fauna as well as flora), and the illustration of the cowslip in *The Loves of the Plants* bears a close resemblance to the one published in that first issue of the *Botanical Magazine* just a few years earlier. Much like Bartram, Darwin wished "to inlist Imagination under the banner of Science," yet such conscription did not extend, it seems, to visual representation.[5]

Erasmus Darwin's views on animal economy and generation, published in *Zoonomia*, bear some overlap with Bartram's vital materialist leanings and have often been compared to that of the French naturalist Jean-Baptiste Lamarck. According to Lamarck, organisms adapted to suit their environments, and these adaptations were then inherited by their offspring. His suggestion that species would acquire new characteristics over an individual's lifetime (rather than through natural selection, as argued by Charles Darwin) resonates in some respects with William Bartram's own study of nature. Yet the images of the natural world found in Lamarck's publications are largely doctrinaire; aside from the plates that portray polyps—those physically and conceptually instable creatures—the book's illustrations hew to the principles of transparency and self-evidence.

Perhaps the visual rhetoric that comes closest to Bartram's own is to be found in Alexander von Humboldt's *Essai sur la géographie des plantes* (1805). Humboldt, a wealthy Prussian polymath, was and remains a giant in the natural sciences, celebrated both during and after his lifetime for examining the patterns and processes of natural plant distribution. Unlike Darwin and Lamarck, Humboldt conducted most of his research in the field, exploring regions of Central and South America, with a special focus on the Amazonian rainforest and the Andean mountains. Through his research, he endeavored to shift the study of botany from the taxonomic description of individual plants to the study of where and how they live. The intransitive and irreducible properties of place that informed Bartram's worldview were a guiding principle of Humboldt's natural history.

To capture the scope of his research in biogeography, Humboldt featured an illustration in his *Essai* that—like Bartram's most inventive work—is a jostling together of different visual strategies. Titled *Géographie des plantes équinoxiales: Tableau physique des Andes et Pays voisins* (fig. C.1), the central figure of this plate is a schematic rendering of Chimborazo and Cotopaxi conjoined (they are in reality at least eighty miles apart), with one half of Chimborazo covered in

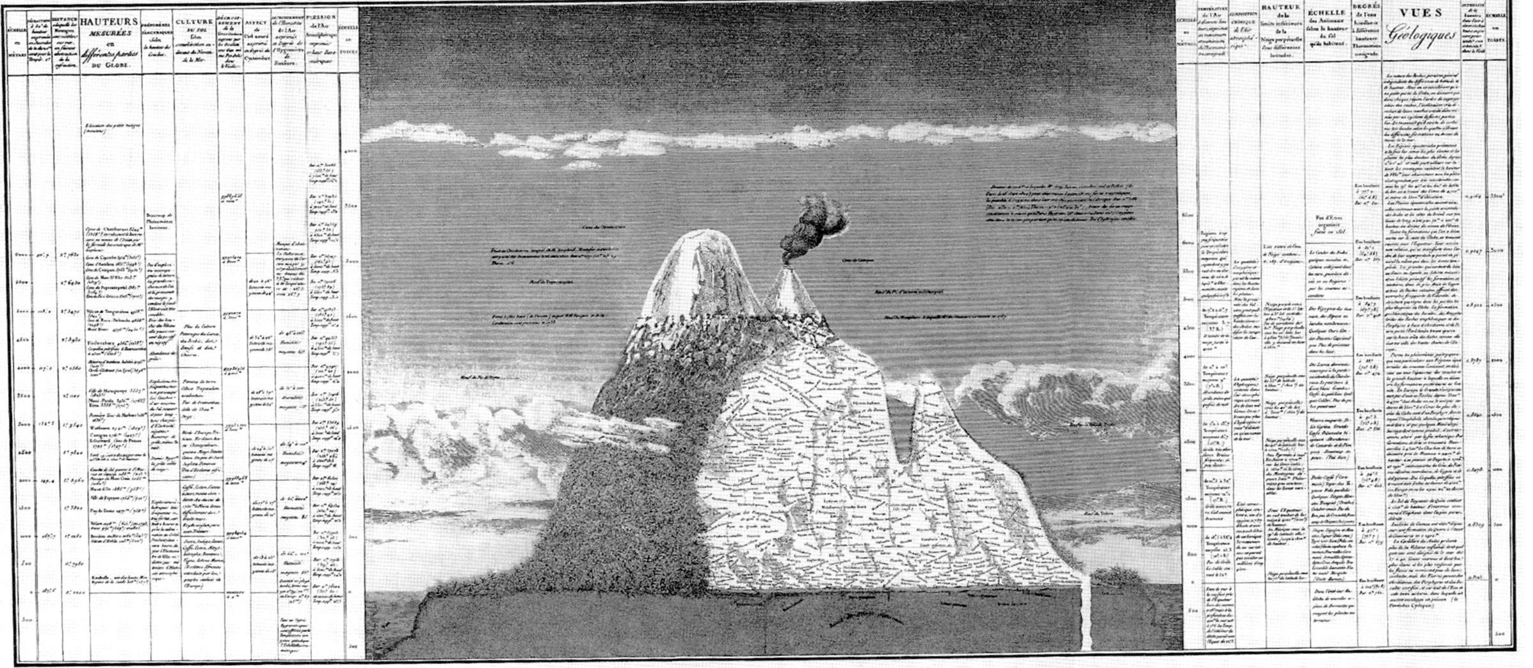

Fig. C.1. Alexander von Humboldt, Lorenz Adolph Schönberger, and Pierre François Turpin (L. Bouquet, engraver), *Tableau physique des Andes et Pays voisins*.

From *Essai sur la géographie des plantes* (1805). © The Board of Trustees of the Royal Botanic Gardens, Kew.

scrubby foliage that grows lighter and sparer as the viewer imaginatively ascends its face. On Chimborazo's other half, which bleeds into Cotopaxi, the visual representation of foliage is replaced instead by plant names to illustrate the altitudes at which they grow. The interlacing of text and image, along with the strange depiction of space—a flatness that is disrupted by the evocation of depth in the overlapping clouds—forces the eye to circulate and integrate these competing representational modes. With the names of Humboldt and his collaborator, Aimé Bonpland, clearly identified along its bottom border, the plate also calls attention to Humboldt's rigorous empirical investigation and the literal heights to which he ascended in order to study the distribution of plant species by latitude and altitude.[6]

In these particulars, Humboldt's *Tableau* seems deeply Bartramian. At the same time, Humboldt explained his desire to balance "two conflicting interests, appearance and exactitude," which is where the two artists perhaps diverge.[7] Humboldt traveled with the most advanced scientific instruments of his day, and he used detailed columns of data to bracket the central figure of Chimborazo and Cotopaxi, visually reinforcing the idea that standardized scientific measurement, as much as empirical experience, formed the basis for his research. Strict measurement was hardly Bartram's concern. His interest in microscopy underscored the relativity of size and scale, and his assessments of size or distance were often intended to evoke emotional or imaginative states rather than to provide hard data.[8]

Despite the resonances between Bartram and Darwin, Lamarck, or Humboldt, there appear to be no direct avenues of influence among their visual rhetoric. There were areas of social and intellectual overlap, certainly: Benjamin Smith Barton liberally cited Erasmus Darwin's *Loves of the Plants* in his *Elements of Botany*, and the specimens Bartram sent to the Jardin du Roi in Paris would have been closely studied by Lamarck. Indeed, in order to describe what he called the *Gordonia pubescens* in his in his *Encyclopédie méthodique*, Lamarck was looking at the very specimens of *Franklinia* that Bartram had provided. Alexander von Humboldt, after his travels in South and Central America, famously visited Philadelphia in 1804, where he was feted at the American Philosophical Society, was introduced to important political and scientific figures, and toured Bartram's Garden.[9]

Although the correspondence is mute on whether Humboldt and Bartram ever met, it would be surprising had they not crossed paths, and the visual representation of nature was a fundamental concern to both. In his five-volume *Cosmos: A Sketch of the Physical Description of the Universe*, published in the mid-nineteenth century, Humboldt pointed to the importance and value of portraying the natural world as a means of drawing attention to it as a subject of study. He encouraged artists to pursue large-scale landscape and panorama painting, since these genres would best evoke that sense of site specificity so crucial to understanding the natural world.[10] The nineteenth-century American landscape painter Frederic Edwin Church would famously take up Humboldt's challenge, yet we already see a concerted effort to convey nature's complex phenomena in Bartram's accretive compositions from nearly a century earlier. In Bartram's graphic enterprise—its whimsy, its curiosity, its lively beauty—we see a naturalist's earnest attempt to craft a new means of understanding and portraying nature, one consistent with the natural world's dynamic and irreducible particularity.

Although William Bartram has been relegated to a coda of eighteenth-century botanical and zoological illustration, his graphic work exemplifies a captivating method of coming to know the natural world, as well as a means of conveying that very knowledge. In this sense, he offered an alternative form of representational transparency to the type that predominated in the eighteenth century and after, a form that did not efface his own exploratory methods but traced them, inviting his viewers to join him on his investigations of the natural world. He focused not on visual standardization but on curiosity and visual wonder, making drawings that often coalesce and dissolve in a process that appears continuous with nature itself.

Notes

Introduction

1. Some of the contemporary scholars studying Bartram's *Travels* often incorporate discussion of his graphic work as well, even if the emphasis remains on the text; see, for instance, Monique Allewaert, *Ariel's Ecology: Plantations, Personhood, and Colonialism in the American Tropics*, 29–82; Christopher Iannini, *Fatal Revolutions: Natural History, West Indian Slavery, and the Routes of American Literature*, 177–217; and Thomas Hallock, "'On the Borders of a New World': Ecology, Frontier Plots, and Imperial Elegy in William Bartram's *Travels*."

2. J. Bartram to Collinson, Sept. 25, 1755, in *The Correspondence of John Bartram: 1734–77*, ed. Edmund Berkeley and Dorothy Smith Berkeley (hereafter cited as *CJB* with page numbers), 387.

3. For the contours of Bartram's life and the history of Bartram's Garden, I am indebted to the work of many contemporary scholars, including the late Joel Fry, the renowned historian of Bartram's Garden; Judith Magee, formerly of the Natural History Museum, London; and Amy Meyers, the director emeritus of the Yale Center for British Art. Early publications by naturalist Francis Harper and historian of science Joseph Ewan remain essential to gaining a sense of Bartram's natural history; their work has been greatly expanded by the Bartram Trail Conference and its 1979 publication, *Bartram Heritage: A Study of the Life of William Bartram*, and by Thomas Hallock and Nancy Hoffmann's edited volume *William Bartram: The Search for Nature's Design* (hereafter cited as *SND*, with page numbers), which compiles previously unpublished primary sources and scholarly essays on Bartram and his milieu.

4. The drawing's annotations are in John Bartram's hand, and scholars have debated whether the drawing itself was made by William or by John.

5. J. Bartram to Peter Collinson, Sept. 25, Apr. 27, 1755, *CJB*, 387, 385. See also the introduction to Hallock and Hoffmann, *SND*, 1–15.

6. Little is known of Captain Child, aside from what can be gathered from the Bartram correspondence and newspaper advertisements. During William's apprenticeship from 1756 to 1760, Child's store was located on Water Street, just north of Market, and dealt primarily in textiles. Around the time Bartram's apprenticeship ended, Child went into business with Richard Stiles—a merchant and slave trader from Bermuda—with whom he sold sugar, molasses, and rum. Their partnership ended in 1762, when Stiles returned to Bermuda; Child died the following year. See *Pennsylvania Gazette*, Mar. 27, 1760; *Pennsylvania Journal*, Nov. 18, 1762; *Pennsylvania Gazette*, Oct. 27, 1763. For Stiles's transport and sale of slaves, see *Pennsylvania Journal*, May 28, 1752; *Pennsylvania Gazette*, May 28, June 11, 1752.

7. W. Bartram to J. Bartram, May 20, 1761, *SND*, 37.

8. J. Bartram to W. Bartram, Sept. 1, Dec. 27, 1761, *SND*, 41, 46.

9. W. Bartram to Isaac Bartram, [1764], *SND*, 48.

10. Collinson to J. Bartram, n.d. [Feb. 3, 1767], *CJB*, 680.

11. Henry Laurens to J. Bartram, Aug. 9, 1766, *SND*, 62. Iannini reads Bartram's "pilgrimage" in *Travels* as an act of penance for the sin of slaveholding. See Iannini, *Fatal Revolutions*, 181. After William's plantation failed, he stayed on in East Florida, where he worked with the cartographer John William Gerard DeBrahm, returned to Philadelphia in 1767, was back in North Carolina by 1770, and returned to Philadelphia briefly before setting off on his botanizing travels for Fothergill. See William Cahill, "William Bartram and the Romance of Learning: A Study in Eighteenth-Century American Education," 98–99.

12. Bartram and the publishers also developed a "deluxe" edition of the book that featured an additional seven hand-colored quarto illustrations. See William Cahill, Joel Fry, Nancy Hoffman, and Alina Josan, "Bartram's *Travels* 1791: A Bibliographic Census," in Kathryn H. Braund, ed., *The Attention of a Traveller: Essays on William Bartram's* Travels *and Legacy*, 244–45.

13. Samuel Johnson, "figure, v.a.," *A Dictionary of the English Language*, 1755, https://johnsonsdictionaryonline.com/1755/figure_va.

14. Laura Rigal, "An American Manufactory: Political Economy, Collectivity, and the Arts in Philadelphia, 1790–1810," 23.

15. Fothergill's response from Collinson to W. Bartram, July 18, 1768, *SND*, 76.

16. Joel Fry notes that despite—or because of—his dedication to documenting Bartram's life and work, Harper often made assumptions that do not always accord with the historical record. Personal correspondence, July 2, 2022.

17. Joseph Ewan's *William Bartram: Botanical and Zoological Drawings, 1756–1788* was the first comprehensive catalogue of Bartram's graphic work at the Natural History Museum, London. Judith Magee drew upon and greatly expanded this project in *The Art and Science of William Bartram*. Magee's book still focuses on the works at the Natural History Museum; no published survey exists of Bartram's complete oeuvre.

18. Amy Meyers, "Sketches from the Wilderness: Changing Conceptions of Nature in American Natural History Illustration, 1680–1880." For her discussion of Bartram's drawings, see esp. chapters 3 and 4.

19. Michael Gaudio, "Swallowing the Evidence: William Bartram and the Limits of Enlightenment," 13.

20. Nancy Hoffmann, "The Construction of William Bartram's Narrative Natural History: A Genetic Text of the Draft Manuscript for *Travels through North and South Carolina, Georgia, East & West Florida*."

21. J. Bartram to Collinson, Dec. 16, 1754, *CJB*, 376. Some draft fragments of John Bartram's arboreal taxonomy exist in the papers of the Historical Society of Pennsylvania, but the whole document is unlocated.

22. Collinson to J. Bartram, Feb. 13, 1753/54, *CJB*, 369.

23. Collinson to J. Bartram, Feb. 13, 1753/54, *CJB*, 369; J. Bartram to Collinson, Nov. 3, 1754, *CJB*, 375.

24. Materialism and vitalism are often presented as opposites in the scholarly literature—the former as simply a mechanistic understanding of living bodies and the latter attributing life to an immaterial vital force or soul. As Charles T. Wolfe has pointed out, the distinction was hardly so rigid. The opposition Wolfe identifies is less between materialism and vitalism and more between mechanistic determinism and organic unpredictability. See Wolfe, "Vitalism and the Resistance to Experimentation on Life in the Eighteenth Century," 269n.26; Wolfe, "Vital Materialism and the Problem of Ethics in the Radical Enlightenment." I use Wolfe's term "vital materialism" throughout to refer to this view of living, animate matter. Jane Bennett has famously examined the notion of vital materialism in a contemporary context; see her *Vibrant Matter: A Political Ecology of Things* (Durham: Duke University Press, 2010).

25. Henry Baker, *The Microscope Made Easy*, xii–xiii.

26. John Turberville Needham, "A Summary of Some Late Observations upon the Generation, Composition, and Decomposition of Animal and Vegetable Substances," *Philosophical Transactions* 45 (1748): 644.

27. Caroline van Eck, *Art, Agency, and Living Presence: From the Animated Image to the Excessive Object*, 18.

28. W. Bartram to R. Barclay, Nov. 1788, *SND*, 148–49.

29. See, for instance, Meyers, "Sketches from the Wilderness"; Larry Clarke, "The Quaker Background of William Bartram's View of Nature"; Cahill, "William Bartram and the Romance of Learning." Kerry S. Walters speaks to the various philosophical strands that Bartram braids together in his study of nature. See Walters, "The Creator's Boundless Palace: William Bartram's Philosophy of Nature"; Walters, "The 'Peaceable Disposition' of Animals: William Bartram on the Moral Sensibility of Brute Creation."

30. Sharece Blakney, "Equally Free with Myself," in *Stories We Know*, ed. Aislinn Pentecost-Farren; Joel Fry, "Slavery and Freedom at Bartram's Garden"; Kerry Walters, "All Equally Dear to God: William Bartram's Antislavery Manuscript," *SND*, 372–76; Allewaert, *Ariel's Ecology*, 29–82; Iannini, *Fatal Revolutions*, 177–217.

31. J. Bartram to W. Bartram, Apr. 5, 1766, *SND*, 55.

32. George Bartram to W. Bartram, Sept. 27, 1773, *SND*, 97. His brother-in-law notes that he sold her to "Captain Masson," who transported her to Charleston, likely to sell again. "Captain Masson" likely refers to the merchant Thomas Mason, who traveled regularly between Philadelphia and Charleston and was a partner in Mason & Hartley, later Mason & Patton. See, for instance, *Pennsylvania Gazette*, Feb. 17, 1773.

33. William Bartram, *Travels Through North & South Carolina, Georgia, East & West Florida . . .* (1791), 312.

34. Trustees of the Public Libraries, North Carolina, *North Carolina Wills and Inventories*, 469–71 (471, 470, 470).

35. William Bartram, "Antislavery Treatise," *SND*, 379–80.

36. W. Bartram to Mary Bartram Robeson, Sept. 7, 1788, *SND*, 141.

Chapter 1: The Asymmetry of Transatlantic Natural History

1. John Bartram to Alexander Catcott, May 26, 1742, *The Correspondence of John Bartram: 1734–77*, ed. Edmund Berkeley and Dorothy Smith Berkeley, *CJB*, 194.

2. Susan Scott Parrish examines the interdependence of the American colonial and the European virtuoso in chapter 3 of *American Curiosity: Cultures of Natural History in the Colonial British Atlantic World*, 103–35.

3. For details on John Bartram's biography, see the editors' introduction to *CJB*, xi–xv; Thomas P. Slaughter, *The Natures of John and William Bartram*; Whitfield J. Bell Jr., "John Bartram: A Biographical Sketch," in Nancy E. Hoffman and John C. Van Horne, eds., *America's Curious Botanist: A Tercentennial Reappraisal of John Bartram, 1699–1777*, 3–22.

4. Library Company of Philadelphia, *"At the Instance of Ben Franklin": A Brief History of the Library Company*, 6; editors' introduction, *CJB*, xii–xiii.

5. Peter Collinson to J. Bartram, Feb. 17, 1737/1738, *CJB*, 84.

6. J. Bartram to Collinson, Dec. 10, 1738, *CJB*, 104–5.

7. The front matter of Hoffman and Van Horne, *America's Curious Botanist*, includes maps and dates of John Bartram's botanizing trips; see xviii–xxvii.

8. J. Bartram to Jan Frederik Gronovius, Dec. 16, 1754, *CJB*, 377.

9. J. Bartram to Collinson, [fall 1753], *CJB*, 364.

10. Collinson to J. Bartram, Aug. 10, 1753, *CJB*, 351.

11. J. Bartram to Gronovius, Dec. 16, 1754, *CJB*, 377.

12. Collinson to J. Bartram, Aug. 10, 1753, Feb. 18, 1756, *CJB*, 351, 397.

13. Collinson to J. Bartram, July 10, 1738, *CJB*, 93.

14. Parrish, *American Curiosity*, 141–44. The author directs special attention to the relationship between Collinson and John Bartram in chapter 4.

15. Collinson to J. Bartram, Dec. 14, 1737, *CJB*, 70.

16. In a letter dated Sept. 20, 1737, Collinson writes that Logan asked him to purchase a copy of "Parkinsons Herbal" for John; see *CJB*, 34. John's volumes of Miller's *Gardeners Dictionary* were gifts from John Jacob Dillenius and Robert James, Lord Petre.

17. Collinson would soon come to recognize John Bartram as an intellectual equal, but he began their correspondence under the assumption that Bartram was an uneducated colonial.

18. Collinson to J. Bartram, Jan. 24, 1734/1735, *CJB*, 4.

19. J. Bartram to Collinson, Apr. 27, 1755, *CJB*, 384.

20. J. Bartram to Collinson, May 1738, *CJB*, 89.

21. William H. Goetzmann, "John Bartram's Journey to Onandaga in Context," in Hoffmann and Van Horne, *America's Curious Botanist*, 97–105.

22. Preface, in John Bartram, *Observations on the Inhabitants, Climate, Soil, Rivers, Productions, Animals, and Other Matters Worthy of Notice . . .*, i.

23. "A letter from John Bartram, M.D. to Peter Collinson, F.R.S., concerning a cluster

of small teeth observed by him at the root of each fang or great tooth in the head of a rattle-snake, upon dissecting it," *Philosophical Transactions* 41 (Jan. 1740): 358–59; "Extract of a letter from Dr. John Bartram to Mr. Peter Collinson, F.R.S., containing some observation concerning the salt-marsh muscle, the oyster-banks, and the fresh-water muscle, of Pensylvania," *Philosophical Transactions* 43 (Jan. 1744): 157–59.

24. Peter Collinson, "Some observations on the *Cicada* of North America, collected by Mr. P. Collinson, F.R.S.," *Philosophical Transactions* 54 (Jan. 1764): 67.

25. Peter Collinson, "An account of some very curious wasps nests made of clay in Pensilvania," *Philosophical Transactions* 43 (Jan. 1744): 363.

26. Peter Collinson, "Some Observations on the Dragon-Fly or Libella of Pensilvania, Collected from Mr. John Bartram's Letters, Communicated by Peter Collinson, F.R.S.," *Philosophical Transactions* 46 (1750): 323–326. Despite the paper's title, Collinson offers his own observations on the European insect. René-Antoine Ferchault de Réaumur's discussion of the libella can be found in volume 6 of his *Mémoires pour servir à l'histoire des insectes* (Paris, 1742), 387–456.

27. "A Further Account of the Libellae or May-flies, from Mr. John Bartram of Pensylvania, Communicated by Mr. Peter Collinson, F.R.S." *Philosophical Transactions* 46 (1749): 400, also 400–402. Parrish notes how Bartram's fragmented epistolary style "was so like its material referent—the sprawled miscellany of American nature"; see Parrish, *American Curiosity*, 142.

28. Collinson to J. Bartram, Feb. 18, 1756, *CJB*, 399.

29. Peter Collinson, "Description of the Miscellaneous Plate," *Gentleman's Magazine* 28 (Jan. 1758): 8–9.

30. Collinson to J. Bartram, Feb. 18, 1756, *CJB*, 399.

31. George Edwards, *Gleanings of Natural History* (London, 1760), 2:192.

32. Edwards, *Gleanings*, 2:198–99.

33. Edwards, *Gleanings*, 2:174.

34. Edwards, *Gleanings*, 2:174.

35. Doherty in chapter 4 discusses the reuse of previously published images as the illustrations for Francis Willughby's *Ornithology* (London, 1678). Doherty argues that the effect of accuracy in this instance was established through the naturalist's process of collecting, collating, and assessing available visual and textual materials; see Meghan Doherty, "Carving Knowledge: Printed Images, Accuracy, and the Early Royal Society of London," 133–73.

36. I would like to thank Tiffany Vogel Biedermann for her assistance in interpreting Bartram's drawing and Edwards's etching.

37. Collinson to J. Bartram, May 28, 1766, *CJB*, 666.

38. Collinson to J. Bartram, n.d. [Feb. 3, 1767], *CJB*, 680. John Bartram's report was included without his knowledge in the second edition of William Stork's *A Description of East-Florida*, published by the Board of Trade and Plantations.

39. J. Bartram to Mark Catesby, Mar. 1740/1741, *CJB*, 152.

40. Collinson to J. Bartram, Sept. 1, 1741, *CJB*, 167. Stephanie Volmer examines at

length the tension between geographical rootedness and mobility in eighteenth-century botanical exchanges; see "Planting a New World: Letters and Languages of Transatlantic Botanical Exchange, 1733–1777" (PhD diss., Rutgers University, 2008), 47–57.

41. Collinson to J. Bartram, Apr. 12, 1739, *CJB*, 118. See also Stephanie Volmer, "Planting a New World," 47.

42. Collinson to J. Bartram, Apr. 24, 1751, *CJB*, 323.

43. Collinson to J. Bartram, Feb. 3, 1741/1742, *CJB*, 180–81.

44. Collinson to J. Bartram, Feb. 3, 1741/1742, *CJB*, 181.

45. J. Bartram to Collinson, [1759], *CJB*, 451–54; Collinson to J. Bartram, July 20, 1759, *CJB*, 469–70.

46. Collinson to J. Bartram, June 30, 1764, *CJB*, 633.

47. Collinson to J. Bartram, May 2, 1738, 91.

48. Robert McCracken Peck, "Preserving Nature for Study and Display," in Prince, *Stuffing Birds, Pressing Plants, Shaping Knowledge*, 15.

49. George Edwards, *A Natural History of Uncommon Birds*, 3:110.

50. J. Bartram to Collinson, [fall 1753], *CJB*, 357. Some rattlesnakes do change color under stress, though it is unclear whether those native to the American Northeast (timber rattlesnakes, or *Crotalus horridus*) do. See John Stepanek, Natalie M. Claunch, Julius A. Frazier, Ignacio T. Moore, Ben J. Vernasco, Camilo Escallón, and Emily N. Taylor, "Corticosterone and Color Change in Southern Pacific Rattlesnakes (*Crotalus helleri*)."

51. Henrietta McBurney, *Illuminating Natural History: The Art and Science of Mark Catesby*, 129.

52. Although Linnaeus conceded that his descriptions of classes and orders were artificial, he believed his genera and species to be natural; see Staffan Müller-Wille and Karen Reeds, "A Translation of Carl Linnaeus's Introduction to *Genera Plantarum* (1737)," 565. In his introduction to a reprint of Carl Linnaeus's *Species Plantarum*, William Stearn points out that Linnaeus attempted to develop a more natural system for classes and orders as early as 1738; Carl Linnaeus, *Species Plantarum*, 1:34.

53. J. Bartram to Philip Miller, Apr. 20, 1755, *CJB*, 380.

54. J. Bartram to Collinson, Aug. 20, 1753, *CJB*, 353–54.

55. Collinson to J. Bartram, Feb. 13, 1753/1754, *CJB*, 369, 372n1.

56. J. Bartram to Collinson, Nov. 3, 1754, *CJB*, 376. I would like to thank Joel Fry, who clarified Bartram's meaning in this passage.

57. Collinson to J. Bartram, Feb. 18, 1756, *CJB*, 399.

58. Philip Miller, *Figures of the Most Beautiful, Useful and Uncommon Plants*, 1:5–6.

Chapter 2: William's Inimitable Picture

An early version of this chapter was first published as "William Bartram's Inimitable Picture: Representation as the Pursuit of Natural Knowledge."

1. J. Bartram to Collinson, Sept. 25, 1755, *CJB*, 387.

2. Collinson to J. Bartram, July 31, 1767, *CJB*, 685.

3. Collinson to W. Bartram, Feb. 16, 1768, *SND*, 72.

4. Collinson to J. Bartram, Feb. 16, 1768, *CJB*, 697.

5. Charlotte Porter views the strangeness of Bartram's work as a result of his colonial status; see her essay "The Drawings of William Bartram (1739–1823), American Naturalist." Christopher Iannini, on the other hand, overlooks the drawing's fragmented quality and suggests instead that it conveys a "sense of epistemological stability and transparency"; Iannini, *Fatal Revolutions*, 179.

6. John Woodward, *Brief Instructions for Making Observations in all Parts of the World*, 12.

7. Collinson to J. Bartram, Jan. 24, 1734/1735, *CJB*, 4.

8. Woodward, *Brief Instructions*, 12. All italics within quotes are in the original text unless otherwise noted.

9. Darrell Sewell, ed., *Philadelphia: Three Centuries of American Art*, exhibition catalogue, 37–39; James Green and Peter Stallybrass, *Benjamin Franklin: Writer and Printer*, 52–57; *The Art of Drawing, and Painting in Water-colours*, 17.

10. Green and Stallybrass, *Benjamin Franklin*, 52–57.

11. See Jennifer L. Roberts, "The Veins of Pennsylvania: Benjamin Franklin's Nature-Print Currency."

12. See Doherty, "Carving Knowledge," 1–11. As Doherty points out, accuracy is a social construction that depends on accepted standards of truth. Although visual representations are of course mediated knowledge, adherence to particular conventions gives them the *effect* of transparency and therefore accuracy.

13. Mark Catesby, *Natural History of Carolina, Florida, and the Bahama Islands*, 1:vi–vii.

14. Catesby, *Natural History*, 3rd ed., 2:115. From the specimens and seeds John provided, Catesby had hoped he might cultivate the tree in Britain. I thank Joel Fry for pointing out the inaccuracy of the depicted flower.

15. Robert Hooke, *Micrographia, or, Some Physiological Descriptions of Minute Bodies Made*, preface, unpaginated.

16. Henry Baker, *Micrographia restaurata* (1745), preface.

17. Meghan C. Doherty, "Discovering the 'True Form': Hooke's *Micrographia* and the Visual Vocabulary of Engraved Portraits." Hooke describes the process of assessing the "true form" in his preface to *Micrographia*. See also Janice Neri, "Between Observation and Image," in Neri, *The Insect in the Image: Visualizing Nature in Early Modern Europe, 1500–1700*, 105–38.

18. Hooke, *Micrographia*, preface.

19. Neri, *Insect in the Image*, 113.

20. Baker, *Micrographia restaurata*, 33.

21. Lorraine Daston and Peter Galison, *Objectivity*, 58.

22. Daston and Galison, *Objectivity*, 55–114.

23. Müller-Wille and Reeds, "Translation of Carl Linnaeus's Introduction," 568.

24. Carl Linnaeus, *Hortus Cliffortianus*, 14. *Collinsonia* was not included in the first edition of *Genera Plantarum* but is described in the second; see Carl Linnaeus, *Genera Plantarum* (1742), 15.

25. Although the root was considered a specific characteristic, Linnaeus preferred when possible to rely on visible parts of the plant for identification. See Kärin Nickelsen, *Draughtsmen, Botanists, and Nature*, 78.

26. This plate was likely not of Ehret's design; his name is usually affixed to his work in the *Hortus Cliffortianus.*

27. Daston and Galison, *Objectivity*, 58–60.

28. Nickelsen, *Draughtsmen, Botanists, and Nature*, 4–5.

29. Edwards, *Natural History*, 1:xix.

30. Edwards, *Natural History*, 1:36. Amy Meyers argues that the unusual juxtapositions in Edwards's illustrations underscore his position as an armchair naturalist. See "Sketches from the Wilderness," 87–88n137.

31. Although the Old World and American lotuses are now considered distinct species (*Nelumbo nucifera* and *Nelumbo lutea*, respectively), I group them together as one, in accordance with eighteenth-century botanical knowledge.

32. Dioscorides describes the plant as growing in abundance in Egypt, Asia, and Sicily; see *De Materia Medica*, 248. For more of the history of the American lotus specifically and its propagation by the Bartrams, see Ewan, *William Bartram*, 60; Joel T. Fry, "The Pond at Bartram's Garden."

33. Parkinson's *Theatrum botanicum* was also in the Bartram family library, a gift from Collinson.

34. John Parkinson, *Theatrum botanicum*, 375. See also Ilaria Maria Grimaldi, Sureshkumar Muthukumaran, Giulia Tozzi, Antonino Nastasi, Nicole Boivin, Peter J. Matthews, and Tinde van Andel, "Literary Evidence for Taro in the Ancient Mediterranean: A Chronology of Names and Uses in a Multilingual World."

35. Parkinson, *Theatrum botanicum*, 375.

36. Parkinson, *Theatrum botanicum*, 377.

37. Pietro Andrea Mattioli, *I discorsi di M. Pietro Andrea Matthioli . . . nelli sei libri di Pedacio Dioscoride Anazarbeo della materia Medicinale*, 448–49.

38. Parkinson, *Theatrum botanicum*, 375.

39. Carolus Clusius, *Exoticorum libri decem*, 32–33; Parkinson, *Theatrum botanicum*, 375–76.

40. William Aiton, in *Hortus Kewensis* (1789), 2:227, identifies the plant's introduction to Britain as 1787, yet a letter dated June 1785 from Margaret Bentinck, Duchess of Portland, to Joseph Banks suggests an earlier date. See section 15, series 72.135, Papers of Sir Joseph Banks, https://bankspapers.sl.nsw.gov.au/banks/section-15/series-72/72-135-letter-received-by-banks-from-the.html

41. Miller's cross-reference is noted in Ewan, *Botanical and Zoological Drawings*, 60.

42. Collinson to J. Bartram, Feb. 22, 1750, *CJB*, 307.

43. N. Bryllion Fagin offers a comprehensive overview of Bartram's influence on literature in *William Bartram: Interpreter of the American Landscape*, 128–94. For Bartram's influence on Samuel Coleridge specifically, see John Livingston Lowes, *The Road to Xanadu: A Study in the Ways of the Imagination*.

44. William Bartram, *Travels Through North & South Carolina, Georgia, East & West Florida*, 98–99. All references to this title are to the 1791 edition.

45. Bartram, *Travels*, 408–9.

46. Noël-Antoine Pluche, *Spectacle de la Nature, or, Nature Display'd*, 5:56–57. Pluche's multivolume work, first published in French in 1732, was translated and widely reprinted. The Bartrams surely owned an English-language copy; in a letter to Sir Hans Sloane, dated Sept. 23, 1743, John Bartram mentions Collinson's gift of "them fine books of Nature Delineated." See William Darlington, *Memorials of John Bartram and Humphry Marshall*, 304. Robert McCracken Peck has traced the location of a number of books owned by John and William in his essay "Shelving Knowledge in Philadelphia: John and William Bartram's Books," in Braund, *Attention of a Traveller*, 219–31, but much of the family's library remains unlocated.

47. Pluche, *Spectacle de la Nature*, 5:54–55.

48. Bartram, *Travels*, 166.

49. Gaudio, "Swallowing the Evidence," 6–7. The quote Gaudio cites from Bartram's undated treatise "The Dignity of Human Nature," can be found in *SND*, 349.

50. Gaudio, "Swallowing the Evidence," 6–7.

51. Bartram, *Travels*, 166.

52. Writing on Copley's *Boy with a Squirrel* (1766; Museum of Fine Arts, Boston), Jennifer Roberts argues that Copley consciously attempts to re-create this very act of synthesis. The painting would have been doubly problematic for Bartram, in that it portrays sensory synthesis without being able to accomplish it. See Jennifer Roberts, *Transporting Visions: The Movement of Images in Early America*, 34–49.

53. As far as I am aware, no other scholar has recognized the heron as a quotation from Edwards.

54. Joshua Kirby, *Dr. Brook Taylor's Method of Perspective Made Easy*, vi.

55. Collinson to J. Bartram, July 31, 1767, *CJB*, 685.

56. William Hogarth, *The Analysis of Beauty*, 87.

57. Daniel Webb, *Inquiry into the Beauties of Painting*, 136–37.

58. Hogarth, *Analysis*, 86–87.

59. Meyers identifies Bartram's consistent use of reflected form and links it to his environmental view of nature; see "Sketches from the Wilderness," 132–38.

60. Colleen Terry, "Presence in Print: William Hogarth in British North America," see esp. 243–49.

61. Leonard Larabee, ed., *Papers of Benjamin Franklin, January 1 through December 31, 1766*, 13:271n4; *The Charter, Laws, and Catalogue of Books of the Library Company of Philadelphia* (1765), 43.

62. Hogarth, *Analysis*, 65.

63. Paulson offers a thorough analysis of Hogarth's rebuke in his introduction; see Hogarth, *Analysis*, xix–xxvi.

64. John Barrell, *Political Theory of Painting from Reynolds to Hazlitt*, 1–68.

65. The primary purpose of Hogarth's book is to teach readers "to *see with [their] own eyes*"; see Hogarth, *Analysis*, 18. Joel Fry notes that there was briefly a drawing master on faculty at the Academy of Philadelphia during William's tenure there, but whether he took lessons is unclear. See Fry, "'To See the Moveing Pensil; Display a Sort of Paper Creation, Which May Endure for Ages': William Bartram as a Natural History Artist," in Braund, *Attention of a Traveller*, 122.

66. Hogarth, *Analysis*, 32. To telegraph the importance of pursuit and discovery, Hogarth featured Christopher Columbus on the subscription ticket for the *Analysis*.

67. Frédéric Ogée, "Je-sais-quoi: William Hogarth and the Representation of the Forms of Life," in Bindman, Ogée, and Wagner, *Hogarth*, 71–84.

68. Hogarth, *Analysis*, 44, 46.

69. Hogarth, *Analysis*, 44, 46.

70. Ogée, "Je-sais-quoi," in Bindman, Ogée, and Wagner, *Hogarth*, 75.

71. Ogée, "Je-sais-quoi," in Bindman, Ogée, and Wagner, *Hogarth*, 74.

72. The authors in the edited volume *Hogarth: Representing Nature's Machines* speak directly to the intersection of the *Analysis* and natural history, whereas other scholars have focused on its relationship to, and influence on, codes of decorum, British nationalism, and ideas of originality. See Terry, "Presence in Print," 245–46n430.

73. Collinson to W. Bartram, Feb. 16, 1768, *SND*, 70–71.

Chapter 3: Vital Matters

1. Collinson to W. Bartram, July 18, 1768, *SND*, 75–76.

2. Fothergill to J. Bartram, n.d., *CJB*, 750.

3. Fothergill to W. Bartram, Oct. 22, 1772, *SND*, 84.

4. Ann Thomson, *Bodies of Thought: Science, Religion, and the Soul in the Early Enlightenment*; see especially chapter 6 for the taxonomic and theological concerns spurred by the concept of vital matter.

5. *Pennsylvania Gazette*, Aug. 2, 16, 23, 1744.

6. The *Pennsylvania Packet*, Mar. 30, 1772; *Dunlap's Pennsylvania Packet, or the General Advertiser*, June 27, July 25, 1774.

7. Leo Lemay, *The Life of Benjamin Franklin, Printer and Publisher, 1730–1747*, 2:110.

8. Henry Baker, *The Microscope Made Easy*, xiv; George Adams, *Micrographia illustrata*, 5.

9. J. Bartram to Col. William Byrd, summer 1739, and to Collinson, Nov. 3, 1754, *CJB*, 120, 375. In a letter to Collinson dated Nov. 13, 1758, *CJB*, 441, he noted that he had just received his copy of Baker. For the popularity of optical displays in early America, see Wendy Bellion, *Citizen Spectator: Art, Illusion, and Visual Perception in Early National America*, 23–62.

10. Baker, *Microscope Made Easy*, 286.

11. Baker, *Micrographia restaurata*, 19.

12. Baker, *Micrographia restaurata*, 19–20.

13. Adams, *Micrographia illustrata*, 232–33.

14. Bartram, *Travels*, 89, 101.

15. Bartram, *Travels*, 133.

16. Bartram, *Travels*, xxix–xxxi.

17. Bartram, *Travels*, xxvi.

18. Bartram, *Travels*, xxvi.

19. Bartram, *Travels*, xxx.

20. Baker, *Micrographia restaurata*, 53–54.

21. For a geologic study of the site, see Michael Ritter, "A Hydrogeology and Origin of the Alachua Sink"; for a historical and cultural analysis, see Henry A. Baker, "Spanish Ranching and the Alachua Sink Site: A Preliminary Report"; Carl Webber, *The Eden of the South*, 20, 28–31.

22. Bartram, *Travels*, 189.

23. Bartram, *Travels*, 199, 195.

24. Bartram, *Travels*, 203.

25. Lucia Nuti, "The Perspective Plan in the Sixteenth Century: The Invention of a Representational Language," 109.

26. Svetlana Alpers, *The Art of Describing: Dutch Art in the Seventeenth Century*, 139–42.

27. Hallock, "On the Borders of a New World," 112. Hallock's interpretation of *The Great Alachua-Savana* derives, as almost all scholars' interpretations do, from the original in-depth analysis in Meyers, "Sketches from the Wilderness," 131–47.

28. Nuti, "Perspective Plan in the Sixteenth Century," 108.

29. Bartram, *Travels*, 20, 94, 155, 132, 190.

30. Baker, *Microscope Made Easy*, 293–94.

31. Bartram, "On the Dignity of Human Nature," *SND*, 354–55, 299.

32. Cuscowilla is present-day Micanopy, Florida. See Francis Harper's annotations for William Bartram, "Travels in Georgia and Florida, 1773–74: A Report to John Fothergill," 184.

33. Ann Thomson points out that the tendency in the history of science to categorize mechanism as wholly dualistic is an oversimplification and that "discussions of vital matter fed apparently contradictory strands of thought, and the same works could be used to support contrasting views." Thomson, *Bodies of Thought*, 67.

34. Guido Giglioni, "What Ever Happened to Francis Glisson? Albrecht Haller and the Fate of Eighteenth-Century Irritability," 465.

35. See John Henry, "Robert Hooke, The Incongruous Mechanist," in *Robert Hooke: New Studies*, ed. Michael Hunter and Simon Schaffer, 149–80.

36. Hooke, *Micrographia*, 212, 123.

37. Wolfe, "Vital Materialism." See also Peter Hanns Reill, *Vitalizing Nature in the Enlightenment*. Catherine Packham offers an overview of vitalism in a British context in *Eighteenth-Century Vitalism: Bodies, Culture, Politics*.

38. For a brief synopsis of Haller's experiments, see Anne C. Vila, *Enlightenment and Pathology: Sensibility in the Literature and Medicine of Eighteenth-Century France*, 13–42; Barbara Maria Stafford, *Body Criticism: Imaging the Unseen in Enlightenment Art and Medicine*, 405–9.

39. Trembley quoted in Virginia P. Dawson, *Nature's Enigma: The Problem of the Polyp in the Letters of Bonnet, Trembley, and Réaumur*, 96–97.

40. Leeuwenhoek published his observations in *Philosophical Transactions* 23 (1702–1703): 1304–11, 1430–43. De Jussieu did not publish on the polyp, though he did send drawings to Réaumur. See Janelle A. Schwartz, "Worm Work: Eighteenth-Century Natural History and Romantic Aesthetics Frontiers," 43–44n10.

41. Susannah Gibson, "On Being Animal, or, the Eighteenth-Century Zoophyte Controversy in Britain," 456–57. See also Gibson's dissertation, "The Pursuit of Nature: Defining Natural Histories in Eighteenth-Century Britain," 76–79.

42. Dawson, *Nature's Enigma*, 101.

43. Dawson, *Nature's Enigma*, 101.

44. Dawson, *Nature's Enigma*, 100–109.

45. Abraham Trembley, "Observations and Experiments upon the Freshwater Polypus, by Monsieur Trembley, at the Hague," v.

46. Trembley, "Observations and Experiments," x. Also see Dawson, *Nature's Enigma*, 124–26, for a description of his method of inverting the polyp.

47. "La Figure, qui remplit toute cette Planche, représente une partie d'un morceau de bois, que j'ai trouvé entre plusieurs autres dans le mois de Juillet 1742, dans un fossé de Sorgvliet. Ce morceau de bois est couvert de Polypes à longs bras." See Abraham Trembley, *Mémoires pour servir à l'histoire d'un genre de polypes d'eau douce*, 226.

48. Trembley, "Observations and Experiments," xi.

49. Mark Ratcliff, *The Quest for the Invisible: Microscopy in the Enlightenment*, 105.

50. This work was also among the microscopy texts available at the Library Company of Philadelphia; see *Charter, Laws, and Catalogue of Books* (1765), 61.

51. Henry Baker, *An Attempt Towards a Natural History of the Polype*, 201, 203.

52. John Turberville Needham, "A Summary of Some Late Observations upon the Generation, Composition, and Decomposition of Animal and Vegetable Substances," 636–37.

53. Needham, "A Summary," 638.

54. Needham, "A Summary," 638, 644.

55. "Part of a Letter from ——— of Cambridge, to a Friend of the Royal Society Occasioned by What Has Lately Been Reported concerning the Insect Mentioned in Page 218 of this Transaction," *Philosophical Transactions* 42 (1742/1743): 229.

56. Bartram, *Travels*, xviii–xx.

57. Bartram, *Travels*, xx.

58. Arthur Dobbs to Peter Collinson, Apr. 2, 1759, Linnaean Society of London. For a comprehensive history of the Venus flytrap and its introduction to Europe, see E. Charles Nelson, *Aphrodite's Mousetrap*; Tim S. Bailey, *Miraculum Naturae—Venus's Flytrap*.

59. Magee, *Art and Science of William Bartram*, 49.

60. Collinson to J. Bartram, May 10, 1763, *CJB*, 593.

61. See Collinson to W. Bartram, July 18, 1768, *SND*, 76.

62. John Ellis, Notebook 2, fols. 104–6, John Ellis Manuscripts and Correspondence, Archives of the Linnean Society of London.

63. Ellis to David Skene, Sept. 24, 1768, quoted in Bailey, *Miraculum Naturae*, 180.

64. John Ellis, *Directions for Bringing Over Seeds and Plants, From the East Indies*, 37.

65. Carl Linnaeus, *Systema Vegetabilium*, 402. In a published letter, Ellis said he felt obligated "to prevent this extraordinary Doctrine . . . from spreading in this Kingdom." See John Ellis, "Curious Experiments on the Seeds of Fungi," *Gentleman's Magazine* 43 (July 1773): 316.

66. Bartram, *Travels*, xx–xxi.

67. Thomas Percival, "Speculations on the Perceptive Power of Vegetables," 13.

68. Meyers, "Sketches from the Wilderness," 159.

69. Alexander Nemerov, *The Body of Raphaelle Peale: Still Life and Selfhood, 1812–1824*, 16.

70. Bartram quoted from Ewan, *Botanical and Zoological Drawings*, 54.

71. Baker, *Attempt Towards a Natural History*, 209.

72. Bartram, *Travels*, xxiv.

73. Fothergill to J. Bartram, Jan. 13, 1770, CJB, 729.

74. Baker, *Microscope Made Easy*, 301.

75. Baker, *Attempt Towards a Natural History*, 211–12.

76. J. Bartram to Fothergill, Nov. 26, 1769, CJB, 723.

77. Daniel Cottom, *Cannibals and Philosophers: Bodies of Enlightenment*, 22–24.

78. Lester C. Olson, *Emblems of American Community in the Revolutionary Era*, 42–43. The segmented snake was originally introduced by Benjamin Franklin to encourage colonial unity during the French and Indian War but was later used to represent resistance to the British.

79. Cottom, *Cannibals and Philosophers*, 22–23.

80. J. Bartram to Fothergill, Nov. 26, 1769, CJB, 723.

81. Bayle quoted in Cottom, *Cannibals and Philosophers*, 22–23.

82. Pierre Bayle, *The Dictionary Historical and Critical of Mr. Peter Bayle* (1737), 4:439.

83. Bayle, *Dictionary Historical and Critical* (1737), 4:431.

84. Catesby, *Natural History*, 1:14.

85. As Meyers observes, Catesby recognized the disruption to local ecologies caused by transplantation. With his etching of the rice birds, he paradoxically depicted a stable natural order, even as he described its potential mutability. See Meyers, "Picturing a World in Flux," in Amy Meyers and Margaret Beck Pritchard, eds., *Empire's Nature: Mark Catesby's New World Vision*, 242–48.

86. Meyers, "Sketches from the Wilderness," 172–77.

87. Meyers, "Sketches from the Wilderness," 181.

88. Cottom, *Cannibals and Philosophers*, 135.

89. Baker, *Attempt toward a Natural History*, 46–47.

90. For more background on Bartram's education at the Academy of Philadelphia, see Cahill, "Romance of Learning," 113–19. Bartram attended the academy sporadically from January 1752 to mid of 1755, or from age twelve to age sixteen.

91. For more information on the curriculum of the Latin School specifically, see Memorandum, Curriculum for the Freshman and Junior Classes in College and in the Latin School, 1791[?], document 1670, Archives General Collection of the University of Pennsylvania, 1740–1820. See also Isaac Anderson Pennypacker, "Alexander Graydon at Pennsylvania." G. Gabrielle Starr discusses the cultural valence of Ovid in her essay "Burney, Ovid, and the Value of the Beautiful."

92. Fothergill to William Huddesford, 10 June [1769?], folios 227–28, MS Ashmole 1822, Bodleian Library, Oxford University.

93. E. J. Kenney, "Introduction," in Ovid, *Metamorphoses*, xix.

94. Ovid, *Metamorphoses*, 177–78.

95. My interpretation of the legs draws on Marcus Wood's discussion of J. M. W. Turner's *Slaver Throwing Overboard the Dead and the Dying, Typhon Coming On* (1840; Museum of Fine Arts, Boston), in which he links the leg in the painting's foreground to Pieter Bruegel's *Landscape with the Fall of Icarus* (ca. 1558; Royal Museum of Fine Arts, Brussels). See Marcus Wood, *Blind Memory: Visual Representations of Slavery in England and America, 1780–1865*, 47–49.

96. Ovid, *Metamorphoses*, 177.

97. Bartram, *Travels*, 167.

Chapter 4: A World in Figures

A shortened version of this chapter was published as "Lively Pictures: William Bartram's Drawing ad vivum," in Braund, *Attention of a Traveller*, 138–51.

1. Claudia Swan, "*Ad vivum, naer het leven*, from the Life: Defining a Mode of Representation." More recently, see the introduction to Balfe, Woodall, and Zittel, *Ad vivum? Visual Materials and the Vocabulary of Life-Likeness in Europe before 1800*, 1–31, as well as Robert Felfe's essay in the same volume, "*Naer het leven*: Between Image-Generating Techniques and Aesthetic Mediation," esp. 72–77.

2. Felfe, "*Naer het leven*," in Balfe, Woodall, and Zittel, *Ad vivum?*, 72.

3. Van Eck, *Art, Agency, and Living Presence*, 18; 45–47.

4. David Rosand, *Drawing Acts: Studies in Graphic Expression and Representation*, 1–23. Rosand develops themes introduced by Maurice Merleau-Ponty in the essay "Eye and Mind."

5. Felfe assesses the importance of direct physical contact in images from the life. See his essay "*Naer het leven*," 81.

6. An indexical representation is one in which its meaning derives from a physical relationship with the object, like a tire track or a gunshot.

7. Pliny the Elder, *The Natural History*, 257.

8. Hogarth, *Analysis*, 11–12. I use "serpentine line" as the more general term for both Hogarth's waving Line of Beauty and his coiling Line of Grace.

9. Hogarth, *Analysis*, 55.

10. I am indebted to Rigal's description of the serpentine line "as the graphic seam at which writing and drawing, narrative and image meet." Rigal, "American Manufactory," 23.

11. See J. Bartram to Collinson, [Dec. 1739] and [Sept. 1743], *CJB*, 130, 224; Gronovius to J. Bartram, June 2, 1746, *CJB*, 278; J. Bartram to Cadwallader Colden, Aug. 16, 1747, *CJB*, 289.

12. Nick Wrightson, "'[Those with] Great Abilities Have Not Always the Best Information': How Franklin's Transatlantic Book-Trade and Scientific Networks Interacted, ca. 1730–1757," 100–101.

13. Adam Casdin, "Before Imagination: Literary Reverie's Opening to the Present," 1–41.

14. J. Hector St. John de Crèvecoeur, *Letters from an American Farmer*, 265.

15. George Bartram (1735–1777), an émigré to Pennsylvania, wrote to family in Scotland to request a copy of their coat of arms for the Bartrams at Kingsessing; see George Bartram to John Bartram (of Carswell, Scotland), Mar. 1, 1760, GD5/593, National Records of Scotland, Edinburgh. George Bartram would become William's brother-in-law upon marrying his sister Ann in 1764.

16. Rosand, *Drawing Acts*, 2.

17. Hogarth, *Analysis*, 42.

18. Script was taught by a method of rote copying; see Jennifer Monaghan, "Writing Instruction," in *Learning to Read and Write in Colonial America*, 273–301.

19. Hogarth, *Analysis*, 33–34.

20. Christoph Irmscher offers a comprehensive discussion of the rattlesnake in the Anglo-American imaginary; see *The Poetics of Natural History: From John Bartram to William James*, 149–87.

21. J. [Joseph] Breintal [Breintnall], "A Letter from Mr. J. Breintal to Mr. Peter Collinson, F.R.S., Containing an Account of What He Felt after Being Bit by a Rattle-Snake," *Philosophical Transactions* 44 (1746–1747): 145.

22. See Bartram, *Travels*, 196; Catesby, *Natural History*, 2:54.

23. Bartram, *Travels*, 270, 196.

24. Bartram, *Travels*, 219–20.

25. Merleau-Ponty, "Eye and Mind," 143–44.

26. The drawing has always been reproduced with a horizontal orientation. See W. Bartram, "Report to John Fothergill," plate 23; Ewan, *Botanical and Zoological Drawings*, 30; Magee, *Art and Science of William Bartram*, 225.

27. Hogarth, *Analysis*, 102.

28. Hogarth, *Analysis*, 103, 111, 109.

29. Van Eck, *Art, Agency, and Living Presence*, 18.

30. Hogarth, *Analysis*, 18. See also van Eck, *Art, Agency, and Living Presence*, 31–34.

31. Thomas Carlyle to Ralph Waldo Emerson, July 8, 1851, reprinted in Charles Eliot Norton, ed., *The Correspondence of Thomas Carlyle and Ralph Waldo Emerson, 1834–1872*, 2:198.

32. Bartram, *Travels*, 118. As Kathryn Braund and others have made clear, Bartram's observations on alligators derived from two separate trips along the St. Johns River that are

condensed in one "fictive uninterrupted journey." See Braund, "Wrestling with Bartram's Alligators," in Braund, *Attention of a Traveller*, 45.

33. Bartram, *Travels*, 118, 119, 122.

34. Bartram, *Travels*, 122–23.

35. Michael Gaudio offers this interpretation in "Swallowing the Evidence," 9–10.

36. Gaudio, "Swallowing the Evidence."

37. Bartram, *Travels*, 206–7.

38. Van Eck, *Art, Agency, and Living Presence*, 68–72.

39. Joseph Addison, "Pleasures of the Imagination," 6:64.

40. Van Eck observes such ambivalence and ambiguities also in the work of Giovanni Battista Piranesi, a contemporary of Hogarth's; see Van Eck, *Art, Agency, and Living Presence*, 146.

41. Bartram, *Travels*, 123.

42. In our conversations, Dorinda Dallmeyer has suggested that this visual passage may be Bartram's attempt to render cypress knees, parts of the root system of some swamp-dwelling trees that project above the water level.

43. Barbara Johnson discusses a similar tactic in Henry David Thoreau's *Walden* (1854), though she uses the literary term "catachresis" rather than *nominum nudum*. Thoreau may well have derived this tactic from Bartram, whose *Travels* inspired his own engagement with nature. See "A Hound, a Bay Horse, and a Turtle Dove: Obscurity in Walden," in Barbara Johnson, *A World of Difference*, 49–56. See also Iannini, *Fatal Revolutions*, 181.

44. Bartram, *Travels*, 238.

45. Cesare Ripa included an alligator in the emblem for "Luxury"; see *Iconologia*, 50. For its inclusion in cabinets of curiosity, see Giuseppe Olmi, "From the Marvellous to the Commonplace: Notes on Natural History Museums," 241.

46. Hans Sloane, *A Voyage to the Islands of Madera, Barbadoes, Nieves, St. Christophers, and Jamaica, with the Natural History . . . of the Last of those Islands* (1725), 2:346–47.

47. Pliny the Elder, *The Natural History*, 251. Elizabeth Wright outlines interpretations of the Zeuxis story by Ernst Gombrich and Jacques Lacan in *Psychoanalytic Criticism: A Reappraisal*, 1879–80.

48. Van Eck, *Art, Agency, and Living Presence*, 59.

49. Baker, *Micrographia restaurata*, 19; for quote, see Baker, *Microscope Made Easy*, 136.

50. John Banister to [Martin Lister?], [1685?], folios 3–3v, Sloane Mss. 3321, British Library, quoted in Raymond Stearns, *Science in the British Colonies of North America*, 205.

51. Ewan, *Botanical and Zoological Drawings*, 50.

52. See Francis Bacon's *Novum Organon*, 185.

53. Carl Linnaeus, *Critica Botanica*, 196–97.

54. Carl Linnaeus, *De Peloria* (1744), quoted in Åke Gustafsson, "Linnaeus' *Peloria*: The History of a Monster," 242.

55. "Quaenam mutatae in Pelorian Linariae causa sit, nos adhuc fugit," Carl Linnaeus, *De Peloria* (1744), 15. Translation by Mike Bortscheller.

56. See the conclusion to Carl Linnaeus, *De Peloria* (1744), 14–17.

57. Gustafsson, "Linnaeus' *Peloria*," 244.

58. Aaron Kennedy, Nisse Goldberg, and Andrew Minnis, "*Exobasidium ferrugineae* sp. nov., Associated with Hypertrophied Flowers of *Lyonia ferruginea*." Affected and unaffected flowers can coexist on a single branch, as in Bartram's drawing.

59. Daniel Worster, *Nature's Economy: A History of Ecological Ideas*, 2–55.

60. Meyers, "Sketches from the Wilderness," 113, 131–39.

61. Linnaeus, *Critica Botanica*, 177.

62. Cynthia Sundberg Wall, *The Prose of Things: Transformations of Description in the Eighteenth Century*, 1–40, 108–13, 214–20, 231–36.

63. Joshua Reynolds, "Fourth Discourse" (1771), in *The Discourses of Joshua Reynolds*, 36. See also Wall, *Prose of Things*, 32–33.

64. Reynolds, "Fourth Discourse," 39.

65. Hogarth quoted from Derek Jarrett, *England in the Age of Hogarth* (New Haven: Yale University Press, 1982, repr. 1992), 169.

66. Anne Puetz, "Drawing from Fancy: The Intersection of Art and Design in Mid Eighteenth-Century London."

67. Hogarth, *Analysis*, 60.

68. Phillip Sloan, "The Buffon-Linnaeus Controversy," 371.

69. Gaudio, "Swallowing the Evidence," 3.

70. John Miller's *An Illustration of the Sexual System of Linnaeus* is an octavo version of his earlier, folio-sized *Illustratio systematis sexualis Linnaei* (London, 1770–1777).

71. Bartram, *Travels*, xxi.

72. Morrison Heckscher and Leslie Greene Bowman, *American Rococo, 1750–1775: Elegance in Ornament*, 54.

73. Linnaeus, *Genera Plantarum* (1742), 435.

74. Bartram, *Travels*, 153–54.

75. Gaudio, "Swallowing the Evidence," 8–9.

76. Merleau-Ponty, "Eye and Mind," 126, 142.

Chapter 5: Refiguring and Recollecting

1. Joel Fry, "An International Catalogue of North American Trees and Shrubs: The Bartram Broadside, 1783," 6–9; Magee, *Art and Science of William Bartram*, 119–20.

2. W. Bartram to Benjamin Smith Barton, Mar. 1791, *SND*, 160.

3. R. Barclay to Thomas Parke, Jul. 25, 1785, Jul. 18, 1789, folders 6 and 8, Barclay Letters, 1688–1794, Haverford College Library Special Collections (hereafter cited as "Barclay Letters" with folder number).

4. Jacob M. Price and Leslie Hannah, "Barclay, David (1729–1809)," *Oxford Dictionary of National Biography*, Sept. 23, 2004, https://doi-org.ezproxy.lib.uconn.edu/10.1093/ref:odnb/37150.

5. See, for instance, David Barclay to William Logan, Nov. 8, 1770, folder 13, box 1,

series 2, Logan-Fisher-Fox Family Papers, 1703–1950, Historical Society of Pennsylvania; R. Barclay to Parke, Sept. 28, 1788, Feb. 3, 1789, folder 8, Barclay Letters.

6. R. Barclay to Parke, Dec. 2, 1783, also R. Barclay to Rebecca Barclay, Jul. 3, 1782, both in folder 6, Barclay Letters.

7. R. Barclay to Parke, Aug. 9, 1783, folder 6, Barclay Letters.

8. R. Barclay to Parke, Dec. 2, 1783, folder 6, Barclay Letters.

9. R. Barclay to Parke, June 1, 1785, folder 7, Barclay Letters.

10. Fry, "International Catalogue," 12.

11. William Davis, "Memoir of Robert Barclay, of Bury Hill," 173, 170.

12. R. Barclay to Parke, Feb. 24, 1789[?], folder 7, Barclay Letters. This letter is actually dated 1787, but it is almost certainly 1789. The letter refers to the recent regency crisis in Britain, which occurred between November 1788 and March 1789. The "S. Bush" who served as courier was likely Solomon Bush, who was introduced at the Medical Society of London in early 1789. The society's founder, John Coakley Lettsom, was Fothergill's protégé and a friend of Barclay's. See "Extract of a Letter from a Gentleman in London, dated 9th Feb., 1789," *Independent Gazetteer* (Apr. 14, 1789).

13. W. Bartram to R. Barclay, Nov. 1788, *SND*, 142. In a letter dated June 1, 1785, Barclay had informed Bartram that a catalogue from Kew was in development; see folder 7, Barclay Letters. The first edition of the *Hortus Kewensis* would be published in three volumes in 1789.

14. This served as the closest thing to an official compendium of recognized plants and their names before the institution of the International Code for Botanical Nomenclature in the early twentieth century.

15. Bettina Dietz, "Linnaeus' Restless System: Translation as Textual Engineering in Eighteenth-Century Botany."

16. W. Bartram to R. Barclay, Nov. 1788, *SND*, 149. The Irish botanist John Fraser, who had been traveling in North America, befriended Walter and brought his manuscript to London for publication. Walter died in January 1789, not long after the *Flora Caroliniana* was published.

17. William Curtis, *The Botanical Magazine; or, Flower-Garden Display'd* 1, no. 1 (Feb. 1787): preface; W. Hugh Curtis, *William Curtis, 1746–1799: Fellow of the Linnean Society, Botanist and Entomologist*, 37.

18. Curtis, Preface, *Botanical Magazine* 1 (1787): not paginated.

19. Collinson to Linnaeus, Aug. 5, 1746, in Alan W. Armstrong, ed., *"Forget not Mee & My Garden . . .": Selected Letters, 1725–1768 of Peter Collinson, F.R.S.*, 135–36; Andrea Wulf, *The Brother Gardeners: A Generation of Gentlemen Naturalists and the Birth of an Obsession*, 108–9, 125.

20. Curtis, "Dodecatheon Meadia. Mead's Dodecatheon, or American Cowslip," *Botanical Magazine* 1 (1787): 12.

21. Curtis, Preface, *Botanical Magazine* 1 (1787), not paginated.

22. "Art. X. *The Botanical Magazine, or Flower Garden Displayed . . . Vol. VII*," *British Critic* 3 (Apr. 1794): 424.

23. W. Bartram to R. Barclay, Nov. 1788, *SND*, 147, 144.

24. Bartram, *Travels*, 406.

25. Bartram, *Travels*, 406.

26. W. Bartram to R. Barclay, Nov. 1788, *SND*, 146–47. See also Joel Fry, "William Bartram's *Oenothera grandiflora*: 'The Most Pompous and Brilliant Herbaceous Plant yet Known to Exist,'" in Kathryn H. Braund and Charlotte M. Porter, eds., *Fields of Vision: Essays on the* Travels *of William Bartram*, 183–203.

27. W. Bartram to R. Barclay, Nov. 1788, *SND*, 145.

28. Brent Elliot, "The Artwork of Curtis's Botanical Magazine," 35.

29. W. Bartram to R. Barclay, Nov. 1788, *SND*, 149.

30. W. Bartram to R. Barclay, Nov. 1788, *SND*, 148–49.

31. Siegesbeck on Linnaeus, as cited in William Stearn's preface to Carl Linnaeus, *Species Plantarum*, not paginated.

32. Linnaeus, *Critica Botanica*, 177.

33. Collinson to Linnaeus, May 13, 1739, in Armstrong, *"Forget not Mee & My Garden,"* 72.

34. W. Bartram to R. Barclay, Nov. 1788, *SND*, 148.

35. John Bartram and Francis Harper, "Diary of a Journey through the Carolinas, Georgia, and Florida from July 1, 1765, to April 10, 1766," *Transactions of the American Philosophical Society* 33, no. 1 (Dec. 1942): 31.

36. Bartram, *Travels*, 16.

37. Magee, *Art and Science of William Bartram*, 145–46.

38. W. Bartram to R. Barclay, Nov. 1788, *SND*, 149.

39. W. Bartram to R. Barclay, Nov. 1788, *SND*, 143–44. For a comprehensive history of the *Franklinia* from its discovery through the nineteenth century, see Joel Fry, "Bartram's Tree: *Franklinia alatamaha*," in Braund, *Attention of a Traveller*, 73–113.

40. The loblolly bay requires a greenhouse to overwinter in Britain. I thank Joel Fry for drawing this to my attention.

41. Collinson to J. Bartram, Feb. 3, 1762, *CJB*, 546.

42. Collinson to J. Bartram, Dec. 28, 1765, *CJB*, 657. Exotic plants from America were incredibly valuable in eighteenth-century Britain, and it was not uncommon for thieves to steal them from well-appointed gardens.

43. Wulf, *Brother Gardeners*, 255.

44. John Ellis, "A Copy of a Letter from Mr. John Ellis, Esq., F.R.S. to Dr. Linnaeus, F.R.S. &c with the Figure and Characters of that Elegant American Evergreentree, Called by the Gardeners the Loblolly-Bay," *Philosophical Transactions* 60 (1770): 519, 520.

45. Ellis to Linnaeus, Dec. 28, 1770, quoted in Edmund Berkeley, "The History of the Naming of the Loblolly Bay," 153.

46. W. Bartram and J. Bartram Jr. to Carl Linnaeus *fils*, Aug. 16, 1783, *SND*, 122.

47. Humphry Marshall, *Arbustrum Americanum; or, the American Grove*, 49–50.

48. Marshall to John Coakley Lettsom, Nov. 4, 1788, in Darlington, *Memorials*, 549.

49. Parke to Marshall, June 18, 1786, box 318, series 2, Ferdinand J. Dreer Autograph Collection, Historical Society of Pennsylvania.

50. Barton to W. Bartram, Dec. 13, 1788, *SND*, 152. Barton was in Europe for a little over two years, where he proposed to publish an early draft of Bartram's *Travels* and split the proceeds. Bartram declined. Suffering severe financial straits abroad, Barton borrowed money from Robert Barclay—whom he would never repay—for his return to Philadelphia. See Barton to W. Bartram, Aug. 26, 1787, *SND*, 136–38; R. Barclay to Parke, Jul. 18, Aug. 8, 1789, folder 8, Barclay Letters.

51. See Fry, "International Catalogue," 31. Fry reprints both the English and French editions of the 1783 catalogue.

52. R. Barclay to Parke, Jul. 3, 1787, folder 7, Barclay Letters.

53. R. Barclay to Parke, May 6, 1789, folder 8, Barclay Letters.

54. Jean-Baptiste Lamarck, *Encyclopédie méthodique*, 2:770.

55. Parke to Marshall, May 18, 1789, box 318, series 2, Ferdinand J. Dreer Autograph Collection, Historical Society of Pennsylvania.

56. Banks to Marshall, May 6, 1789, box 10/4, series X, Manuscripts, Humphry Marshall Papers, USDA History Collection, Special Collections, National Agricultural Library.

57. The *Franklinia* requires a greenhouse in Europe, and only very rarely did it pollinate or produce ripe fruit. I am thankful to Joel Fry for bringing this to my attention.

58. Bartram's copy of the *Hortus Kewensis* is now in the collection of the Pennsylvania Horticultural Society.

59. André Michaux, *Flora Boreali-Americana*, 1:103–4. Although Bartram twice mentioned the feverbark in his *Travels*, he alternated between the genera *Bartramia* (xviii) and *Bignonia* (468). This may explain why Michaux's designation of the genus remains, despite its later publication date.

60. "975. Hydrangea Quercifolia. Oak-Leaved Hydrangea," *Curtis's Botanical Magazine* 25 (1807).

61. Aiton, *Hortus Kewensis*, 2:231. See Fry, "Bartram's Tree: *Franklinia alatamaha*," in Braund, *Attention of a Traveller*, 88.

62. Charles Louis L'Héritier de Brutelle, *Stirpes Novae* (Paris, 1791), 156. For the proposed generic name of *Lacthea*, see William Hooker, *Paradisus Londinensis*, lvi. Although the book is attributed to Hooker, its text was written by Richard Anthony Salisbury.

63. Charles Jenkins, "Historical Background of Franklin's Tree," 205–6.

64. W. Bartram to Barton, Nov. 27, 1788, *SND*, 150. See also Nancy Hoffmann, "William Bartram's Draft Manuscript for *Travels*: Private Journal and Public Book," *SND*, 282–94.

65. Washington Irving to James Kirke Paulding, May 27, 1820, quoted in Michael Warner, "Irving's Posterity," 783–84.

66. John Locke, *An Essay Concerning Human Understanding*, 148.

67. Hogarth, *Analysis*, 22. Jerome Mazzaro offers a thorough examination of Hogarth's mnemonics in "The Arts of Memory and William Hogarth's Line of Beauty."

68. W. Bartram to R. Barclay, Nov. 1788, *SND*, 149.

69. Locke, *Essay Concerning Human Understanding*, 149.

70. Bartram, *Travels*, 426.

71. W. Bartram to Barton, Mar. 1791, *SND*, 160. See also Amy Meyers, "From Nature and Memory: William Bartram's Drawings of North American Flora and Fauna," in Meyers, *Knowing Nature*, 133.

72. Benjamin Smith Barton, *Fragments of the Natural History of Pennsylvania* (1883), v.

73. Meyers, "From Nature and Memory," in Meyers, *Knowing Nature*, 133–38. Meyers discusses the relationship between Bartram and Barton, especially as regards the collaborative nature of botanical and zoological inquiry and its inherent incompleteness.

74. Hallock and Hoffman, *SND*, 185.

75. W. Bartram to Barton, Aug. 1801, *SND*, 188. See also Joseph Ewan, "From Calcutta and New Orleans, or, Tales from Barton's Greenhouse," 128.

76. By including the pipe bowl in this drawing, Bartram was likely visually referencing plates from Moses Harris's *The Aurelian* (London, 1766). My thanks go to Amy Meyers and Jim Green for this observation.

77. Ewan, *Botanical and Zoological Drawings*, 64.

78. Benjamin Smith Barton, *The Elements of Botany*, x–xi.

79. He explained that his eyes felt as though they were pierced by a sword, and he could bear no daylight for weeks; see Bartram, *Travels*, 418–20.

80. Barton to W. Bartram, Oct. 24, 1801, *SND*, 190–92; W. Bartram to Barton, Oct. 25, 1801, *SND*, 192.

81. Barton to W. Bartram, Nov. 30, 1805, *SND*, 223; W. Bartram to Thomas Jefferson, Feb. 6, 1806, *SND*, 225.

82. Joseph Dennie, "Account of Bartram's Garden," *The Port Folio* VI:II (August 1818): 151.

83. Many of the later visitors to the garden are mentioned in Hallock and Hoffman, *SND*, 158–59; the appellation "Philosopher of Kingsessing" originated with the German physician Johann Heinrich Ferdinand von Autenrieth. Other visitors to the garden are described in Elizabeth Fairhead's dissertation, "Essential Nature: Bartram's Garden and Natural History in Philadelphia, 1790–1825."

84. Fairhead describes Bartram's Garden as just such a laboratory; see "Essential Nature," 1–28. For George Washington's impression of the garden, see his diary entry from June 10, 1787, in *The Diaries of George Washington*, vol. 5, *1 July 1786–31 December 1789* (1979), 166.

85. Volmer addresses this topic specifically in chapter 5 of her dissertation, "Planting a New World," 221–54.

86. Henry Muhlenberg to William Baldwin, June 18, 1812, in William Darlington, *Reliquiae Baldwinianae*, 63–64.

87. Hallock and Hoffman, *SND*, 231; Magee, *Art and Science of William Bartram*, 196.

88. Charlie Williams, Eliane M. Norman, and Walter Kingsley Taylor, trans. and ed., *André Michaux in North America: Journals and Letters, 1785–1797*, 63.

89. Williams, Norman, and Taylor, *André Michaux in North America*, 112; Thomas Walter, *Flora Caroliniana*, 159.

90. Fry notes that the *Franklinia* is included in Michaux's herbarium, but the specimen's origin is unknown.

91. Quoted in Joseph Ewan and Nesta Ewan, "John Lyon, Nurseryman and Plant Hunter, and His Journal, 1799–1814," 12.

92. John Watson to H. Marshall, Apr. 8, 1788, Humphry and Moses Marshall Papers, 1721–1863, William L. Clements Library, University of Michigan.

93. M. Marshall to J. Banks, Oct. 39, 1790, Humphry and Moses Marshall Papers, 1721–1863, William L. Clements Library, University of Michigan.

94. Ewan and Ewan, "John Lyon," 22–23.

Conclusion

1. Ewan, *Botanical and Zoological Drawings*, 31–33.

2. Daston and Galison, *Objectivity*, 57. As Amy Meyers observes, Mark Catesby and George Edwards were not systematists like Linnaeus and both could be willfully strange in their illustrations; still, I understand their aims were as much to naturalize as to mystify. See Meyers, "Sketches from the Wilderness," 46–112.

3. Erasmus Darwin, *Zoonomia; or, the Laws of Organic Life* (London, 1794), 1:146.

4. Erasmus Darwin, *The Loves of the Plants*, in *The Botanic Garden: A Poem in Two Parts*, part 2, vii, 6.

5. Darwin, *The Economy of Vegetation*, in *The Botanic Garden*, part 1, iii. See also Michael Gaudio, "The Elements of Botanical Art: William Bartram, Benjamin Smith Barton, and the Scientific Imagination," in Hallock and Hoffmann, *SND*, 426–49, in which he discusses Bartram's drawings in relation to Darwin's writing.

6. Sylvie Romanowski, "Humboldt's Pictorial Science: An Analysis of the *Tableau physique des Andes et pays voisins*," in Alexander von Humboldt and Aimé Bonpland, *Essay on the Geography of Plants*, 158, 163–64.

7. Humboldt and Bonpland, *Essay on the Geography of Plants*, 81.

8. Slaughter, *Natures of John and William Bartram*, 190–91. See also Braund, Wrestling with Bartram's Alligators," in Braund, *Attention of a Traveller*, 45.

9. For information on Humboldt's visit to Philadelphia, see Andrea Wulf, *The Invention of Nature: Alexander von Humboldt's New World*, 96–100; Fairhead, "Essential Nature," 89–90.

10. Alexander von Humboldt, *Cosmos: A Sketch of the Physical Description of the Universe*, 2:456–57.

Selected Bibliography

Archival Sources

Archives General Collection of the University of Pennsylvania, Philadelphia, 1740–1820.

MS Ashmole 1822, Bodleian Library, University of Oxford, UK.

Papers of Sir Joseph Banks, State Library of New South Wales, Sydney, Australia.

Barclay Letters, 1688–1794, Haverford College Library Special Collections, Haverford, Pennsylvania.

William Bartram Commonplace Book, ca. 1760–1800, Private Collection.

Papers of the Bertram family of Nisbet, Lanark, 1571–1890, National Records of Scotland, Edinburgh.

John Ellis Manuscripts and Correspondence, Archives of the Linnean Society of London.

Logan-Fisher-Fox Family Papers, 1703–1950, Historical Society of Pennsylvania, Philadelphia.

Humphry Marshall Correspondence, Ferdinand J. Dreer Autograph Collection, Historical Society of Pennsylvania, Philadelphia.

Humphry Marshall Papers, USDA History Collection, Special Collections, National Agricultural Library, Washington, DC.

Humphry and Moses Marshall Papers, 1721–1863, William L. Clements Library, The University of Michigan, Ann Arbor.

Newspapers and Periodicals

The Botanical Magazine; or, Flower-Garden Display'd (London, 1787–1800)

Curtis's Botanical Magazine (London, 1801–1920)

Gentleman's Magazine (London, 1734–1834)

Independent Gazetteer (Philadelphia, 1782–1790)

Pennsylvania Gazette (Philadelphia, 1728–1800)

Pennsylvania Journal (Philadelphia, 1742–1793)

Pennsylvania Packet; and the General Advertiser (Philadelphia, 1771–1784)

Philosophical Transactions of the Royal Society (London, 1665–1887)

Washington Quarterly Magazine of Arts, Science and Literature (Washington, DC, 1823–1824)

Books and Articles

Abrams, M. H. *The Mirror and the Lamp: Romantic Theory and the Critical Tradition.* New York: Oxford University Press, 1953.

Adams, Charles. "Reading Ecologically: Language and Play in Bartram's *Travels.*" *Southern Quarterly* 32, no. 4 (Summer 1994): 65–74.

Adams, George. *Micrographia illustrata.* London, 1746.

Addison, Joseph. "Pleasures of the Imagination." 1712. Reprinted in *The Spectator,* 8th ed. Vol. 6, 58–99. London, 1739.

Aiton, William. *Hortus Kewensis.* 3 vols. London, 1789.

Aiton, William. *Hortus Kewensis.* 2d ed. 5 vols. London, 1810–1813.

Allewaert, Monique. *Ariel's Ecology: Plantations, Personhood, and Colonialism in the American Tropics.* Minneapolis: University of Minnesota Press, 2013.

Allewaert, Monique. "Swamp Sublime: Ecologies of Resistance in the American Plantation Zone." *PMLA* 123, no. 2 (Mar. 2008): 340–57.

Alpers, Svetlana. *The Art of Describing: Dutch Art in the Seventeenth Century.* Chicago: University of Chicago Press, 1983.

Anderson, Douglas. "Bartram's *Travels* and the Politics of Nature." *Early American Literature* 25, no. 1 (1990): 3–17.

Anonymous. *Catalogus Plantarum Horti Medici Oxoniensis.* 1648.

Armstrong, Alan W., ed. *"Forget not Mee & My Garden . . .": Selected Letters, 1725–1768 of Peter Collinson, F.R.S.* Philadelphia: American Philosophical Society, 2002.

The Art of Drawing, and Painting in Water-colours. 4th ed. London, 1735.

Athens, Elizabeth. "Chaotic Life: Representing the Freshwater Polyp." *J18: A Journal of Eighteenth-Century Art and Culture* (Aug. 2016), https://www.journal18.org/nq /chaotic-life-representing-the-freshwater-polyp-by-elizabeth-athens.

Athens, Elizabeth. "The Vital Ornament: Natural Philosophy in William Hogarth's *The Analysis of Beauty* (1753)." *Oxford Art Journal* 40, no. 3 (Dec. 2017): 397–418.

Athens, Elizabeth. "William Bartram's Inimitable Picture: Representation as the Pursuit of Natural Knowledge." *Journal of Florida Studies* 1, no. 4 (Dec. 2015), https://www .journaloffloridastudies.org/files/vol0104/04Athens.pdf.

Bacon, Sir Francis. *Novum Organon.* Translated by G. W. Kitchin. Oxford, 1855.

Bailey, Tim S. *Miraculum Naturae—Venus's Flytrap.* Victoria, BC: Trafford, 2008.

Balfe, Thomas, Joanna Woodall, and Claus Zittel, eds. *Ad vivum? Visual Materials and the Vocabulary of Life-Likeness in Europe before 1800.* Leiden: Brill, 2019.

Baker, Henry. *An Attempt towards a Natural History of the Polype.* London, 1743.

Baker, Henry. *Employment for the Microscope.* London, 1753.

Baker, Henry. *Micrographia restaurata.* London, 1745.

Baker, Henry. *The Microscope Made Easy.* 2d ed. London, 1743.

Baker, Henry A. "Spanish Ranching and the Alachua Sink Site: A Preliminary Report." *Florida Anthropologist* 46, no. 2 (June 1993): 82–101.

Barrell, John. *Political Theory of Painting from Reynolds to Hazlitt*. New Haven, CT: Yale University Press, 1986.

Barton, Benjamin Smith. *The Elements of Botany, or Outlines of the Natural History of Vegetables*. Philadelphia, 1803.

Barton, Benjamin Smith. *The Elements of Botany, or Outlines of the Natural History of Vegetables*. Revised and expanded edition by William P. C. Barton. Philadelphia, 1836.

Barton, Benjamin Smith. *Fragments of the Natural History of Pennsylvania*. 1799. Reprinted, edited, and with an introduction by Osbert Salvin. Cambridge, MA, 1883.

Bartram, John. *Observations on the Inhabitants, Climate, Soil, Rivers, Productions, Animals, and Other Matters Worthy of Notice, Made by Mr. John Bartram, in his Travels from Pensilvania to Onondago, Oswego, and the Lake Ontario, in Canada*. London, 1751.

Bartram, John, and Peter Collinson. "Some Observations on the Dragon-Fly or Libella of Pensilvania, Collected from Mr. John Bartram's Letters, Communicated by Peter Collinson, F.R.S." *Philosophical Transactions* 46 (1749): 323–25.

Bartram, William. "Travels in Georgia and Florida, 1773–74: A Report to John Fothergill." Edited and with an introduction by Francis Harper. *Transactions of the American Philosophical Society* 33, no. 2 (Nov. 1943): 121–242.

Bartram, William. *Travels Through North & South Carolina, Georgia, East & West Florida, the Cherokee Country, the Extensive Territories of the Muscogulges, or Creek Confederacy, and the Country of the Chactaws; Containing An Account of the Soil and Natural Productions of Those Regions, Together with Observations on the Manners of the Indians*. Philadelphia, 1791.

Bartram, William. *Travels Through North & South Carolina, Georgia, East & West Florida, the Cherokee Country, the Extensive Territories of the Muscogulges, or Creek Confederacy, and the Country of the Chactaws; Containing An Account of the Soil and Natural Productions of Those Regions, Together with Observations on the Manners of the Indians*. London, 1792.

Bartram Heritage: A Study of the Life of William Bartram. Montgomery, AL: Bartram Trail Conference, 1979.

Batsaki, Yota, Sarah Cahalan, and Anatole Tchikine, eds. *The Botany of Empire in the Long Eighteenth Century*. Washington, DC: Dumbarton Oaks Research Library and Collection, 2016.

Bayle, Pierre. *The Dictionary Historical and Critical of Mr. Peter Bayle*. 2d ed. 5 vols. Translated by Pierre Desmaizeaux. London, 1734–1738.

Bellion, Wendy. *Citizen Spectator: Art, Illusion, and Visual Perception in Early National America*. Chapel Hill: University of North Carolina Press, 2011.

Bennett, Thomas Peter. "The 1817 Florida Expedition of the Academy of Natural Sciences." *Proceedings of the Academy of Natural Sciences of Philadelphia* 152, no. 1 (Oct. 2002): 1–21.

Berkeley, Edmund. "The History of the Naming of the Loblolly Bay." *Journal of the History of Biology* 3, no. 1 (Spring 1970): 149–54.

Berkeley, Edmund, and Dorothy Smith Berkeley, eds. *The Correspondence of John Bartram, 1734–77.* Tallahassee: University Presses of Florida, 1992.

Berkeley, Edmund, and Dorothy Smith Berkeley. *The Life and Travels of John Bartram from Lake Ontario to the River St. John.* Tallahassee: University Presses of Florida, 1982.

Bermingham, Ann. *Learning to Draw: Studies in a Cultural History of a Polite and Useful Art.* New Haven, CT: Yale University Press, 2000.

Bindman, David, Frédéric Ogée, and Peter Wagner, eds. *Hogarth: Representing Nature's Machines.* Manchester: Manchester University Press, 2001.

Boorstin, Daniel. *The Lost World of Thomas Jefferson.* New York: Henry Holt, 1948.

Braund, Kathryn H., ed. *The Attention of a Traveller: Essays on William Bartram's* Travels *and Legacy.* Tuscaloosa: University of Alabama Press, 2022.

Braund, Kathryn H., and Charlotte M. Porter, eds. *Fields of Vision: Essays on the* Travels *of William Bartram.* Tuscaloosa: University of Alabama Press, 2010.

Braund, Kathryn H., and Gregory Waselkov, eds. *William Bartram on the Southeastern Indians.* Lincoln: University of Nebraska Press, 2002.

Breitwieser, Mitchell Robert. "Jefferson's Prospect." *Prospects* 10 (1985): 315–52.

Burns, Sarah. *Painting the Dark Side: Art and the Gothic Imagination in Nineteenth-Century America.* Berkeley: University of California Press, 2004.

Cahill, Edward. *The Liberty of the Imagination: Aesthetic Theory, Literary Form, and Politics in the Early United States.* Philadelphia: University of Pennsylvania Press, 2012.

Cahill, William. "William Bartram and the Romance of Learning: A Study in Eighteenth-Century American Education." PhD diss., Rutgers University, New Brunswick, New Jersey, 2001.

Casdin, Adam. "Before Imagination: Literary Reverie's Opening to the Present." PhD diss., Stanford University, Stanford, California, 2004.

Cashin, Edward. *William Bartram and the American Revolution on the Southern Frontier.* Columbia: University of South Carolina Press, 2000.

A Catalogue of Books Belonging to the Library Company of Philadelphia. Philadelphia, 1741.

Catesby, Mark. *Natural History of Carolina, Florida, and the Bahama Islands.* 1st edition. 2 vols. London, 1731–1743.

Catesby, Mark. *Natural History of Carolina, Florida, and the Bahama Islands.* 2nd edition. 2 vols. London, 1754.

Catesby, Mark. *Natural History of Carolina, Florida, and the Bahama Islands.* 3rd edition. 2 vols. London, 1771.

Cavanilles, Antonio José. *Sexta dissertatio botanica.* Madrid, 1788.

The Charter, Laws, and Catalogue of Books of the Library Company of Philadelphia. Philadelphia: B. Franklin and D. Hall, 1757, and subsequent editions.

Christie, John R. R. "Ideology and Representation in Eighteenth-Century Natural History." *Oxford Art Journal* 13, no. 1 (1990): 3–10.

Clarke, Larry. "The Quaker Background of William Bartram's View of Nature." *Journal of the History of Ideas* 46, no. 3 (July–Sept. 1985): 435–48.

Clusius, Carolus. *Exoticorum libri decem.* Antwerp, 1605.

Collinson, Peter. "An Account of the First Introduction of American Seeds into Great Britain." 1766. Reprinted with notes by A. B. Rendle. *Journal of Botany* 63 (1925): 163–65.

Cottom, Daniel. *Cannibals and Philosophers: Bodies of Enlightenment.* Baltimore, MD: Johns Hopkins University Press, 2001.

Crary, Jonathan. *Techniques of the Observer.* Cambridge, MA: MIT Press, 1990.

Crèvecoeur, J. Hector St. John de. *Letters from an American Farmer.* 1782. Reprint, New York: Fox, Duffield, 1994.

Curtis, W. Hugh. *William Curtis, 1746–1799: Fellow of the Linnean Society, Botanist and Entomologist.* Winchester, UK: Warren and Son, 1941.

Curtis, William. *Flora Londinensis.* 6 vols. London, 1777–1798.

Darlington, William, ed. *Memorials of John Bartram and Humphry Marshall.* Philadelphia, 1849.

Darlington, William, ed. *Reliquiae Baldwinianae.* Philadelphia, 1843.

Darwin, Erasmus. *The Botanic Garden: A Poem in Two Parts.* 3d ed. London, 1794.

Darwin, Erasmus. *Zoonomia; or, the Laws of Organic Life.* 2 vols. London, 1794–1796.

Daston, Lorraine. "Fear and Loathing of the Imagination in Science." *Daedalus* 127, no. 1 (Winter 1998): 73–95.

Daston, Lorraine, and Peter Galison. *Objectivity.* New York: Zone Books, 2007.

Davis, William. "Memoir of Robert Barclay, of Bury Hill." *Quarterly Magazine and Review Chiefly Designed for the Use of the Society of Friends* (1882): 170–73.

Dawson, Virginia P. *Nature's Enigma: The Problem of the Polyp in the Letters of Bonnet, Trembley, and Réaumur.* Philadelphia: American Philosophical Society, 1987.

De Bolla, Peter. *The Education of the Eye: Painting, Landscape and Architecture in Eighteenth-Century Britain.* Stanford, CA: Stanford University Press, 2003.

De Certeau, Michel. "The Madness of Vision." *Enclitic* 7, no. 1 (Spring 1983): 24–31.

DeFillips, R. "Parasitism in *Ximenia*." *Rhodora* 71, no. 787 (July–Sept. 1969): 439–43.

Dickinson, Victoria. *Drawn from Life: Science and Art in the Portrayal of the New World.* Toronto: University of Toronto Press, 1998.

Dietz, Bettina. "Linnaeus' Restless System: Translation as Textual Engineering in Eighteenth-Century Botany." *Annals of Science* 73, no. 2 (2016): 143–56.

Dioscorides. *De materia medica.* Edited and translated by T. A. Obaldeston and R. P. A. Wood. Johannesburg: Ibidis Press, 2000.

Doherty, Meghan. "Carving Knowledge: Printed Images, Accuracy, and the Early Royal Society of London." PhD diss., University of Wisconsin–Madison, 2010.

Doherty, Meghan. "Discovering the 'True Form': Hooke's *Micrographia* and the Visual Vocabulary of Engraved Portraits." *Notes and Records of the Royal Society* 66 (2012): 211–34.

Edwards, George. *Gleanings of Natural History.* 3 vols. London, 1758–1764.

Edwards, George. *A Natural History of Uncommon Birds.* 4 vols. London, 1743–1751.

Elliot, Brent. "The Artwork of Curtis's Botanical Magazine." *Curtis's Botanical Magazine* 17, no. 1 (Feb. 2000): 35–41.

Ellis, John. *Directions for Bringing Over Seeds and Plants, From the East-Indies, and Other Distant Countries, in a State of Vegetation* . . . London, 1770.

Ellis, John. *An Essay Toward a Natural History of the Corallines.* London, 1755.

Ewan, Joseph. "From Calcutta and New Orleans, or, Tales from Barton's Greenhouse." *Proceedings of the American Philosophical Society* 127, no. 3 (June 1983): 125–34.

Ewan, Joseph. *William Bartram: Botanical and Zoological Drawings, 1756–1788.* Philadelphia: American Philosophical Society, 1968.

Ewan, Joseph, and Nesta Ewan. "John Lyon, Nurseryman and Plant Hunter, and His Journal, 1799–1814." *Transactions of the American Philosophical Society* 53, no. 2 (1963): 1–69.

Fagin, N. Bryllion. *William Bartram: Interpreter of the American Landscape.* Baltimore, MD: Johns Hopkins University Press, 1933.

Fairhead, Elizabeth. "Essential Nature: Bartram's Garden and Natural History in Philadelphia, 1790–1825." PhD diss., Michigan State University, East Lansing, 2005.

Felfe, Robert. "*Naer het leven*: Between Image-Generating Techniques and Aesthetic Mediation." Translated by Jonathan Lutes. In *Ad vivum? Visual Materials and the Vocabulary of Life-Likeness in Europe before 1800*, edited by Thomas Balfe, Joanna Woodall, and Claus Zittel, 44–88. Leiden: Brill, 2019.

Foshay, Ella M. "Charles Darwin and the Development of American Flower Imagery." *Winterthur Portfolio* 15, no. 4 (Winter 1980): 299–314.

Foucault, Michel. *The Order of Things: An Archaeology of the Human Sciences.* 1971. Reprint, New York: Vintage Books, 1994.

Franklin, Benjamin. *Proposals Relating to the Education of Youth in Pennsilvania.* Philadelphia, 1749.

Fry, Joel. "An International Catalogue of North American Trees and Shrubs: The Bartram Broadside, 1783." *Journal of Garden History* 16, no. 1 (1996): 3–66.

Fry, Joel. "The Pond at Bartram's Garden." *Bartram Broadside* (Summer 1997): 1–7.

Fry, Joel. "Slavery and Freedom at Bartram's Garden." Presentation at the conference Investigating Mid-Atlantic Plantations: Slavery, Economies, and Space, Philadelphia, Pennsylvania, Oct. 17, 2019.

Gage, A. T. *A History of the Linnean Society of London.* London: The Linnean Society, 1938.

Gaudio, Michael. "Swallowing the Evidence: William Bartram and the Limits of Enlightenment." *Winterthur Portfolio* 36, no. 1 (Spring 2001): 1–17.

Gell, Alfred. *Art and Agency: An Anthropological Theory.* Oxford: Clarendon Press, 1998.

Gibson, Susannah. "On Being Animal, or, the Eighteenth-Century Zoophyte Controversy in Britain." *History of Science* 50, no. 1 (Dec. 2012): 453–76.

Gibson, Susannah. "The Pursuit of Nature: Defining Natural Histories in Eighteenth-Century Britain." PhD diss., University of Cambridge, UK, 2011.

Giglioni, Guido. "What Ever Happened to Francis Glisson? Albrecht Haller and the Fate of Eighteenth-Century Irritability." *Science in Context* 21, no. 4 (2008): 465–93.

Gustafsson, Åke. "Linnaeus' *Peloria*: The History of a Monster." *Theory of Applied Genetics* 54 (1979): 241–48.

Green, James, and Peter Stallybrass. *Benjamin Franklin: Writer and Printer*. Philadelphia: Library Company of Philadelphia, 2006.

Grimaldi, Ilaria Maria, Sureshkumar Muthukumaran, Giulia Tozzi, Antonino Nastasi, Nicole Boivin, Peter J. Matthews, and Tinde van Andel. "Literary Evidence for Taro in the Ancient Mediterranean: A Chronology of Names and Uses in a Multilingual World." *PLoS ONE* 13, no. 6 (June 2018), doi.org/10.1371/journal.pone.0198333.

Hallock, Thomas. "Male Pleasure and the Genders of Eighteenth-Century Botanic Exchange: A Garden Tour." *William and Mary Quarterly* 62, no. 4 (Oct. 2005): 697–718.

Hallock, Thomas. "'On the Borders of a New World': Ecology, Frontier Plots, and Imperial Elegy in William Bartram's *Travels*." *South Atlantic Review* 66, no. 4 (Autumn 2001): 109–33.

Hallock, Thomas, and Nancy Hoffmann, eds. *William Bartram: The Search for Nature's Design*. Athens: University of Georgia Press, 2010.

Hanson, Craig Ashley. *The English Virtuoso: Art, Medicine, and Antiquarianism in the Age of Empiricism*. Chicago: University of Chicago Press, 2009.

Harper, Francis. "Proposals for Publishing Bartram's *Travels*." *Library Bulletin of the American Philosophical Society* (1945): 27–38.

Harper, Francis, ed. *The Travels of William Bartram, Naturalist Edition*. 1958. Reprint. Athens: University of Georgia Press, 1998.

Harper, Francis. "William Bartram and the American Revolution." *Proceedings of the American Philosophical Society* 97, no. 5 (Oct. 1953): 571–77.

Harper, Francis, and Arthur N. Leeds. "A Supplementary Chapter on *Franklinia Alatamaha*." *Bartonia* no. 19 (1937): 1–13.

Heckscher, Morrison, and Leslie Greene Bowman. *American Rococo, 1750–1775: Elegance in Ornament*. New York: Metropolitan Museum of Art, 1992. Exhibition catalogue.

Heering, Peter. "The Enlightened Microscope: Re-enactment and Analysis of Projections with Eighteenth-Century Solar Microscopes." *British Journal for the History of Science* 41, no. 3 (Sept. 2008): 345–67.

Hermann, Peter. *Paradisus Batavus*. Leiden, 1698.

Hill, John. *Essay in Natural History and Philosophy*. London, 1752.

Hill, John. *Review of the Works of the Royal Society of London*. London, 1751.

Hoffmann, Nancy. "The Construction of William Bartram's Narrative Natural History: A Genetic Text of the Draft Manuscript for *Travels through North and South Carolina, Georgia, East & West Florida*." PhD diss., University of Pennsylvania, Philadelphia, 1996.

Hoffmann, Nancy E., and John C. Van Horne, eds. *America's Curious Botanist: A Tercentennial Reappraisal of John Bartram, 1699–1777*. Philadelphia: The American Philosophical Society, 2004.

Hogarth, William. *The Analysis of Beauty*. 1753. Reprinted with an introduction and notes by Ronald Paulson. New Haven, CT: Yale University Press, 1997.

Hooke, Robert. *Micrographia, or, Some Physiological Descriptions of Minute Bodies Made*. London, 1665.

Hooker, William. *Paradisus Londinensis*. London, 1803.

Humboldt, Alexander von. *Cosmos: A Sketch of the Physical Description of the Universe*. 5 vols. Translated by E. C. Otté. New York: Harper & Brothers, 1849–1858.

Humboldt, Alexander von, and Aimé Bonpland. *Essay on the Geography of Plants*. Edited and with an introduction by Stephen T. Jackson, translated by Sylvie Romanowksi. Chicago: University of Chicago Press, 2009.

Hunter, Michael, and Simon Schaffer, eds. *Robert Hooke: New Studies*. Suffolk, UK: Boydell Press, 1989.

Iannini, Christopher. *Fatal Revolutions: Natural History, West Indian Slavery, and the Routes of American Literature*. Chapel Hill: University of North Carolina Press, 2012.

Irmscher, Christoph. *The Poetics of Natural History: From John Bartram to William James*. New Brunswick, NJ: Rutgers University Press, 1999.

Ivins, William. *Prints and Visual Communication*. Cambridge, MA: Harvard University Press, 1953.

Jenkins, Charles. "The Historical Background of Franklin's Tree." *Pennsylvania Magazine for History and Biography* 58, no. 3 (1933): 193–208.

Johnson, Barbara. *A World of Difference*. Baltimore, MD: Johns Hopkins University Press, 1987.

Kennedy, Aaron, Nisse Goldberg, and Andrew Minnis. "*Exobasidium ferrugineae* sp. nov., Associated with Hypertrophied Flowers of *Lyonia ferruginea*." *Mycotaxon* 120 (Apr.–June 2012): 451–60.

King, F. Wayne. "Alligator Behavior: The Accuracy of William Bartram's Observations." http://ufdc.ufl.edu/UF00088969/00001.

Kirby, Joshua. *Dr. Brook Taylor's Method of Perspective Made Easy*. London, 1754.

Knellwolf, Christa. "Robert Hooke's *Micrographia* and the Aesthetics of Empiricism." *Seventeenth Century* 16, no. 1 (2001): 177–200.

Kumar, Pramod Mishra. "'[A]ll the World was America': The Transatlantic (Post)Coloniality of John Locke, William Bartram, and the Declaration of Independence." *CR: The New Centennial Review* 2, no. 1 (Spring 2002): 213–58.

Lamarck, Jean-Baptiste. *Encyclopédie méthodique*. 8 vols. Paris and Liège, 1783–1808.

Larabee, Leonard, ed. *Papers of Benjamin Franklin, January 1 through December 31, 1766*. Vol. 13. New Haven, CT: Yale University Press, 1969.

Latour, Bruno. "Visualization and Cognition: Drawing Things Together." In *Knowledge and Society: Studies in the Sociology of Culture Past and Present*, edited by H. Kuklick, translated by Bruno Latour, 1–33. Greenwich: Jai Press, 1990.

Lee, Berta Grattan. "William Bartram: Naturalist or 'Poet'?" *Early American Literature* 7 (Fall 1972): 124–29.

Lemay, Leo. *The Life of Benjamin Franklin: Printer and Publisher, 1730–1747*. Vol. 2. Philadelphia: University of Pennsylvania Press, 2006.

Library Company of Philadelphia. *"At the Instance of Ben Franklin": A Brief History of the Library Company*. Philadelphia: The Library Company, 2015.

Linnaeus, Carl. *Critica Botanica*. 1737. Reprint, translated by Sir Arthur Hoyt. London: Royal Society, 1938.

Linnaeus, Carl. *Genera Plantarum*. Leiden, 1737.

Linnaeus, Carl. *Genera Plantarum*. 2d ed. Leiden, 1742.

Linnaeus, Carl. *Genera Plantarum*. 5th ed. Stockholm, 1754.

Linnaeus, Carl. *Hortus Cliffortianus*. Amsterdam, 1737.

Linnaeus, Carl. *Nova Plantarum Genera*. Uppsala, 1751.

Linnaeus, Carl. *Species Plantarum*. 1753. 2 vols. Reprinted with an introduction and notes by William Stearn. London: The Ray Society, 1957–1959.

Linnaeus, Carl. *Systema Naturae*. Leiden, 1735.

Linnaeus, Carl. *Systema Naturae*. 10th ed. 2 vols. Stockholm, 1758.

Linnaeus, Carl. *Systema Vegetabilium*. Edited by Johan Andreas Murray. Göttingen, 1784.

Locke, John. *An Essay Concerning Human Understanding*. 1690. Reprinted with an introduction and notes by Roger Woolhouse. London: Penguin Books, 2004.

Looby, Christopher. "The Constitution of Nature: Taxonomy as Politics in Jefferson, Peale, and Bartram." *Early American Literature* 22, no. 3 (1987): 252–73.

Lowes, John Livingston. *The Road to Xanadu: A Study in the Ways of the Imagination*. Boston: Houghton Mifflin, 1927.

Mader, Rodney. "Selves Evident: Subjectivity and History in Eighteenth-Century America." PhD diss., Temple University, Philadelphia, Pennsylvania, 1998.

Magee, Judith. *The Art and Science of William Bartram*. University Park: Penn State University Press, 2007.

Marshall, Humphry. *Arbustrum Americanum; or, the American Grove*. Philadelphia, 1785.

Mattioli, Pietro Andrea. *I discorsi di M. Pietro Andrea Matthioli . . . nelli sei libri di Pedacio Dioscoride Anazarbeo della materia Medicinale*. Venice, 1568.

Mazzaro, Jerome. "The Arts of Memory and William Hogarth's Line of Beauty." *Essays in Literature* 20, no. 2 (Fall 1993): 213–30.

McBurney, Henrietta. *Illuminating Natural History: The Art and Science of Mark Catesby*. London: Paul Mellon Centre, 2021.

Merleau-Ponty, Maurice. "Eye and Mind." In Maurice Merleau-Ponty, *The Merleau-Ponty Aesthetics Reader*, edited by Galen A. Johnson, translated by Michael B. Smith, 121–49. Evanston, IL: Northwestern University Press, 1993.

Merleau-Ponty, Maurice. *The Merleau-Ponty Aesthetics Reader*. Edited and with an introduction by Galen A. Johnson. Translated by Michael B. Smith. Evanston, IL: Northwestern University Press, 1993.

Meyers, Amy, ed. "Art and Science in America: Issues of Representation." Special issue, *Huntington Library Quarterly* 59, no. 2–3 (1998).

Meyers, Amy, ed. *Knowing Nature: Art and Science in Philadelphia, 1740–1840*. New Haven, CT: Yale University Press, 2011.

Meyers, Amy. "Sketches from the Wilderness: Changing Conceptions of Nature in

American Natural History Illustration, 1680–1880." PhD diss., Yale University, New Haven, Connecticut, 1985.

Meyers, Amy, and Margaret Beck Pritchard, eds. *Empire's Nature: Mark Catesby's New World Vision*. Chapel Hill: University of North Carolina Press, 1998.

Michaux, André. *Flora boreali-americana, sistens caracteres plantarum quas in America septentrionali*. 2 vols. Paris, 1803.

Michaux, François André. *North American Sylva*. Translated by Augustus Hillhouse. 3 vols. Paris, 1819.

Miller, David Philip, and Peter Hanns Reill, eds. *Visions of Empire: Voyages, Botany, and Representations of Nature*. New York: Cambridge University Press, 1996.

Miller, John. *An Illustration of the Sexual System of Linnaeus*. 2 vols. London, 1779.

Miller, John. *Illustratio systematis sexualis Linnaei*. London, 1770–1777.

Miller, Philip. *Figures of the Most Beautiful, Useful and Uncommon Plants*. 2 vols. London, 1760.

Miller, Philip. *The Gardeners Dictionary*. London, 1731 and subsequent editions.

Monaghan, Jennifer. *Learning to Read and Write in Colonial America*. Amherst: University of Massachusetts Press, 2005.

Montanus, Arnold. *Atlas Chinensis: Being a Second Part of a Relation of Remarkable Passages in Two Embassies from the East-India Company of the United Provinces to the Vice-Roy Singlamong*. Translated by John Ogilby. London, 1671.

Moore, Hugh. "The Aesthetic Theory of William Bartram." *Essays in the Arts and Sciences* 12, no. 1 (Mar. 1983): 17–35.

Morison, Robert, and Jacob Bobart. *Plantarum historiae universalis Oxoniensis*. 3 vols. Oxford, 1680–1699.

Müller-Wille, Staffan. "Collection and Collation: Theory and Practice of Linnaean Botany." *Studies in History and Philosophy of Biological and Biomedical Sciences* 38 (2007): 541–62.

Müller-Wille, Staffan, and Karen Reeds. "A Translation of Carl Linnaeus's Introduction to *Genera Plantarum* (1737)." *Studies in History and Philosophy of Biological and Biomedical Sciences* 38 (2007): 563–72.

Needham, John Turberville. *New Microscopical Discoveries*. London, 1745.

Needham, John Turberville. "A Summary of Some Late Observations upon the Generation, Composition, and Decomposition of Animal and Vegetable Substances." *Philosophical Transactions* 45 (1748): 615–66.

Nelson, E. Charles. *Aphrodite's Mousetrap*. Aberystwyth, Wales: Boethius Press, 1991.

"Nelumbium Luteum." *Meehan's Monthly* 10 (Aug. 1900): 113–14.

Nemerov, Alexander. *The Body of Raphaelle Peale: Still Life and Selfhood, 1812–1824*. Berkeley: University of California Press, 2001.

Neri, Janice. *The Insect in the Image: Visualizing Nature in Early Modern Europe, 1500–1700*. Minneapolis: University of Minnesota Press, 2011.

Nickelsen, Kärin. "Draughtsmen, Botanists, and Nature: Constructing Eighteenth-Century Botanical Illustrations." *Studies in History and Philosophy of Biological and Biomedical Sciences* 37, no. 1 (Mar. 2006): 1–25.

Nickelsen, Kärin. *Draughtsmen, Botanists, and Nature: The Construction of Eighteenth-Century Botanical Illustrations*. Dordrecht: Springer, 2006.

Norton, Charles Eliot, ed. *The Correspondence of Thomas Carlyle and Ralph Waldo Emerson, 1834–1872*. 2 vols. Boston, 1884.

Nuti, Lucia. "The Perspective Plan in the Sixteenth Century: The Invention of a Representational Language." *Art Bulletin* 76, no. 1 (Mar. 1994): 105–28.

Nygren, Edward, ed. *Views and Visions: American Landscape before 1830*. Washington, DC: Corcoran Gallery of Art, 1986. Exhibition catalogue.

Ogée, Frédéric, ed. *The Dumb Show: Image and Society in the Works of William Hogarth*. Oxford: Voltaire Foundation, 1997.

Olmi, Giuseppe. "From the Marvellous to the Commonplace: Notes on Natural History Museums." In *Non-verbal Communication in Science prior to 1900*, edited by Renato Mazzolini, 235–78. Florence: L. S. Olschki, 1993.

Olson, Lester C. *Emblems of American Community in the Revolutionary Era*. Washington, DC: Smithsonian Institution Press, 1991.

O'Neill, Jean. *Peter Collinson and the Eighteenth-Century Natural History Exchange*. Philadelphia: American Philosophical Society, 2008.

Otter, Samuel. *Melville's Anatomies*. Berkeley: University of California Press, 1999.

Ovid. *Metamorphoses*. Translated by A. D. Melville, with an introduction by E. J. Kenney. New York: Oxford University Press, 1986.

Packham, Catherine. *Eighteenth-Century Vitalism: Bodies, Culture, Politics*. New York: Palgrave MacMillan, 2012.

Packham, Catherine. "Narrating Resuscitation: Theory, Knowledge, and the Cultural Life of Eighteenth-Century Vitalism." *Literature and Medicine* 37, no. 2 (Fall 2019): 346–67.

Parkinson, John. *Theatrum botanicum*. London, 1640.

Parrish, Susan Scott. *American Curiosity: Cultures of Natural History in the Colonial British Atlantic World*. Chapel Hill: University of North Carolina Press, 2006.

Parsons, James. *The Microscopical Theatre of Seeds*. London, 1745.

Pennypacker, Isaac Anderson. "Alexander Graydon at Pennsylvania." *Alumni Register* (June 1905): 408–12.

Pentecost-Farren, Aislinn, ed. *Stories We Know*. Philadelphia: Bartram's Garden, 2017.

Percival, Thomas. "Speculations on the Perceptive Power of Vegetables." *Memoirs of the Literary and Philosophical Society of Manchester* 2 (Mar. 1785): 1–19.

Pliny the Elder. *The Natural History*. Translated and edited by John Bostock and H. T. Riley. Vol. 6. London: Taylor and Francis, 1855.

Pointon, Marcia. *Naked Authority: The Body in Western Painting, 1830–1908*. Cambridge: Cambridge University Press, 1990.

Pluche, Noël-Antoine. *Spectacle de la Nature: or, Nature Display'd*. 3d ed. 7 vols. Translated by Samuel Humphreys. London, 1740–1753.

Plukenet, Leonard. *Almagestum botanicum*. London, 1696.

Plukenet, Leonard. *Phytographia*. 3 vols. London, 1691–1692.

Porter, Charlotte M. "The Drawings of William Bartram (1739–1823), American Naturalist." *Archives of Natural History* 16, no. 3 (1989): 289–303.

Porter, Charlotte M. "Following Bartram's 'Track': Titian Ramsay Peale's Florida Journey." *Florida Historical Quarterly* 61 (Apr. 1983): 431–44.

Price, Jacob M., and Leslie Hannah. "Barclay, David (1729–1809)." *Oxford Dictionary of National Biography.* Sept. 23, 2004. https://doi-org.ezproxy.lib.uconn.edu/10.1093/ref:odnb/37150.

Prince, Sue Ann, ed. *Stuffing Birds, Pressing Plants, Shaping Knowledge: Natural History in North America, 1730–1860.* Philadelphia: American Philosophical Society, 2003.

Prown, Jonathan, and Richard Miller. "The Rococo, the Grotto, and the Philadelphia High Chest." *American Furniture* 4 (1996): 105–36.

Puetz, Anne. "Drawing from Fancy: The Intersection of Art and Design in Mid-eighteenth-Century London." *RIHA Journal* (Mar. 2014): 1–35.

Ratcliff, Mark. *The Quest for the Invisible: Microscopy in the Enlightenment.* Burlington, VT: Ashgate, 2009.

Ray, John. *Methodus plantarum nova.* London, 1682.

Réaumur, René-Antoine Ferchault de. *Mémoires pour servir à l'histoire des insectes.* 6 vols. Paris, 1734–1742.

Regis, Pamela. *Describing Early America: Bartram, Jefferson, Crèvecoeur, and the Influence of Natural History.* Philadelphia: University of Pennsylvania Press, 1999.

Reill, Peter Hanns. *Vitalizing Nature in the Enlightenment.* Berkeley: University of California Press, 2005.

Reveal, James. "Identification of the Plants and Animals Illustrated by Mark Catesby for His *Natural History of Carolina, Florida, and the Bahama Islands.*" *Phytoneuron* 6 (2013): 1–55.

Reynolds, Joshua. *The Discourses of Joshua Reynolds.* With an introduction by Austin Dobson. London: Oxford University Press, 1907.

Rheede tot Drakenstein, Hendrik Adriaan van. *Hortus Malabaricus.* 12 vols. Amsterdam, 1678–1703.

Rigal, Laura. "An American Manufactory: Political Economy, Collectivity, and the Arts in Philadelphia, 1790–1810." PhD diss., Stanford University, Stanford, California, 1989.

Ripa, Cesare. *Iconologia.* 1593. Translated by P. Tempest. London, 1709.

Riskin, Jessica. *Science in the Age of Sensibility: The Sentimental Empiricists of the French Enlightenment.* Chicago: University of Chicago Press, 2002.

Ritter, Michael. "A Hydrogeology and Origin of the Alachua Sink." MS thesis, University of Florida, Gainsville, 1991.

Roberts, Jennifer L. *Transporting Visions: The Movement of Images in Early America.* Berkeley: University of California Press, 2014.

Roberts, Jennifer L. "The Veins of Pennsylvania: Benjamin Franklin's Nature-Print Currency." *Grey Room* 69 (Fall 2017): 50–79.

Rosand, David. *Drawing Acts: Studies in Graphic Expression and Representation.* Cambridge: Cambridge University Press, 2002.

Schafer, Daniel. *William Bartram and the Ghost Plantations of British East Florida*. Gainesville: University Presses of Florida, 2010.

Schimmelman, Janice. "Books on Drawing and Painting Techniques Available in Eighteenth-Century American Libraries and Bookstores." *Winterthur Portfolio* 19, no. 2/3 (Summer–Autumn 1984): 193–205.

Schmidt, Leigh Eric. *Hearing Things: Religion, Illusion, and the American Enlightenment*. Cambridge, MA: Harvard University Press, 2000.

Schwartz, Janelle A. "Worm Work: Eighteenth-Century Natural History and Romantic Aesthetics Frontiers." PhD diss., University of Winsconsin–Madison, 2008.

Seelye, John. "Beauty Bare: William Bartram and His Triangulated Wilderness." *Prospects* 6 (1981): 37–54.

Sewell, Darrell, ed. *Philadelphia: Three Centuries of American Art*. Philadelphia: Philadelphia Museum of Art, 1976. Exhibition catalogue.

Silver, Bruce. "William Bartram's and Other Eighteenth-Century Accounts of Nature." *Journal of the History of Ideas* 39, no. 4 (Oct.–Dec. 1978): 597–614.

Slaughter, Thomas P. *The Natures of John and William Bartram*. New York: Alfred A. Knopf, 1996.

Sloan, Phillip. "The Buffon-Linnaeus Controversy." *Isis* 67, no. 3 (Sept. 1976): 356–75.

Sloane, Hans. *A Voyage to the Islands of Madera, Barbadoes, Nieves, St. Christophers, and Jamaica, with the Natural History . . . of the Last of those Islands*. 2 vols. London, 1707–1725.

Solkin, David. *Painting for Money: The Visual Arts and the Public Sphere in Eighteenth-Century England*. New Haven, CT: Yale University Press, 1992.

Stabile, Susan. *Memory's Daughters: The Material Culture of Remembrance in Eighteenth-Century America*. Ithaca: Cornell University Press, 2004.

Stafford, Barbara. *Body Criticism: Imaging the Unseen in Enlightenment Art and Medicine*. Cambridge, MA: MIT Press, 1991.

Stafford, Barbara. *Voyage into Substance: Art, Science, Nature, and the Illustrated Travel Account, 1760–1840*. Cambridge: MIT Press, 1984.

Starr, G. Gabrielle. "Burney, Ovid, and the Value of the Beautiful." *Eighteenth-Century Fiction* 24, no. 1 (Fall 2011): 77–104.

Stearns, Raymond. *Science in the British Colonies of North America*. Urbana: University of Illinois Press, 1970.

Stepanek, John, Natalie M. Claunch, Julius A. Frazier, Ignacio T. Moore, Ben J. Vernasco, Camilo Escallón, and Emily N. Taylor. "Corticosterone and Color Change in Southern Pacific Rattlesnakes (*Crotalus helleri*)." *Herpetologica* 75 (2019): 143–52.

Stork, William. *A Description of East-Florida, With A Journal Kept by John Bartram of Philadelphia, Botanist to His Majesty for the Floridas; Upon a Journey from St. Augustine Up the River St. John's, as Far as the Lakes. With Explanatory Botanical Notes*. London, 1769.

Swan, Claudia. "*Ad vivum, naer het leven*, from the Life: Defining a Mode of Representation." *Word & Image* 11, no. 4 (1995): 353–72.

Tatler. Edited by Donald F. Bond. 3 vols. Oxford: Clarendon Press, 1987.

Terry, Colleen. "Presence in Print: William Hogarth in British North America." PhD diss., University of Delaware, Newark, 2014.

Thomson, Ann. *Bodies of Thought: Science, Religion, and the Soul in the Early Enlightenment.* Oxford: Oxford University Press, 2008.

Thornton, Tamara Plakins. *Handwriting in America: A Cultural History.* New Haven, CT: Yale University Press, 1996.

Tournefort, Joseph Pitton de. *Élémens de botanique.* 3 vols. Paris, 1604.

Trembley, Abraham. *Mémoires pour servir à l'histoire d'un genre de polypes d'eau douce.* Leiden, 1744.

Trembley, Abraham. "Observations and Experiments upon the Freshwater Polypus, by Monsieur Trembley, at the Hague." *Philosophical Transactions* 42 (1742/1743): iii–xi.

Trustees of the Public Libraries, Raleigh, North Carolina. *North Carolina Wills and Inventories.* Raleigh, NC: Edwards & Broughton, 1912.

Van Eck, Caroline. *Art, Agency, and Living Presence: From the Animated Image to the Excessive Object.* Leiden: De Gruyter and Leiden University Press, 2020.

Ventenant, E. P. *Jardin de la Malmaison.* 2 vols. Paris, 1803–1804.

Vila, Anne C. *Enlightenment and Pathology: Sensibility in the Literature and Medicine of Eighteenth-Century France.* Baltimore, MD: Johns Hopkins University Press, 1998.

Wall, Cynthia Sundberg. *The Prose of Things: Transformations of Description in the Eighteenth Century.* Chicago: University of Chicago Press, 2006.

Walter, Thomas. *Flora Caroliniana.* London, 1788.

Walters, Kerry S. "The Creator's Boundless Palace: William Bartram's Philosophy of Nature." *Transactions of the Charles S. Peirce Society* 25, no. 3 (Summer 1989): 309–32.

Walters, Kerry S. "The 'Peaceable Disposition' of Animals: William Bartram on the Moral Sensibility of Brute Creation." *Pennsylvania History* 56, no. 3 (July 1989): 157–76.

Warner, Michael. "Irving's Posterity." *ELH* 67, no. 3 (Fall 2000): 773–99.

Washington, George. *The Diaries of George Washington.* Edited by Donald Jackson and Dorothy Twohig. 6 vols. Charlottesville: University Press of Virginia, 1976–1979.

Webb, Daniel. *Inquiry into the Beauties of Painting.* 2d ed. London, 1761.

Webber, Carl. *The Eden of the South.* New York, 1883.

Weiss, Harry, and Grace Ziegler. *Thomas Say: Early American Naturalist.* Baltimore: C. C. Thomas, 1931.

Wells, K. L. H. "Serpentine Sideboards, Hogarth's *Analysis*, and the Beautiful Self." *Eighteenth-Century Studies* 46, no. 3 (2013): 399–413.

Williams, Charlie, Eliane M. Norman, and Walter Kingsley Taylor, trans. and eds. *André Michaux in North America: Journals and Letters, 1785–1797.* Tuscaloosa: University of Alabama Press, 2020.

Wolfe, Charles T. "Vitalism and the Resistance to Experimentation on Life in the Eighteenth Century." *Journal of the History of Biology* 46, no. 2 (Summer 2013): 255–82.

Wolfe, Charles T. "Vital Materialism and the Problem of Ethics in the Radical Enlightenment." *Philosophica* 88 (2013): 31–70.

Wood, Marcus. *Blind Memory: Visual Representations of Slavery in England and America, 1780–1865*. New York: Routledge, 2000.

Woodward, John. *Brief Instructions for Making Observations in all Parts of the World*. London, 1696.

Worster, Daniel. *Nature's Economy: A History of Ecological Ideas*. 2d ed. Cambridge: Cambridge University Press, 1994.

Wright, Elizabeth. *Psychoanalytic Criticism: A Reappraisal*. 2d ed. Cambridge: Polity Press, 1998.

Wrightson, Nick. "'[Those with] Great Abilities I Iave Not Always the Best Information': How Franklin's Transatlantic Book-Trade and Scientific Networks Interacted, ca. 1730–1757." *Early American Studies* 8, no. 1 (Winter 2010): 94–119.

Wulf, Andrea. *The Brother Gardeners: A Generation of Gentlemen Naturalists and the Birth of an Obsession*. New York: Alfred A. Knopf, 2009.

Wulf, Andrea. *The Invention of Nature: Alexander von Humboldt's New World*. New York: Alfred A. Knopf, 2015.

Yokota, Kariann. *Unbecoming British: How Revolutionary America Became a Postcolonial Nation*. New York: Oxford University Press, 2011.

Index

Note: Page numbers in *italics* refer to figures.